FARMING THE SMALL FOREST

A well-managed hardwood forest on the Bazile Tree Farm in Wisconsin, *American Forest Council photo by Lester A. DeCoster.*

Also published by Miller Freeman Publications for the forest products industry:

Operations Management in the Forest Products Industry
By Richard F. Baldwin

Computer Control Systems For Log Processing and Lumber Manufacturing
By Ed M. Williston

Small Log Sawmills: Profitable Product Selection, Process Design and Operation
By Ed M. Williston

Small Log Sawmills: Profitable Product Selection, Process Design and Operation
By Ed M. Williston

Saws: Design, Selection, Operation, Maintenance
By Ed M. Williston

Lumber Manufacturing: The Design and Operation of Sawmills and Planer Mills, Revised Edition
By Ed M. Williston

Dry Land Log Handling & Sorting: Planning, Construction & Operation of Log Yards
By Charles M. Hampton

Modern Particleboard and Dry-Process Fiberboard Manufacturing
By Thomas M. Maloney

Quality Control in Lumber Manufacturing
Edited by Terence D. Brown

Plywood Manufacturing Practices: Revised, Second Edition
By Richard F. Baldwin

Timber Cutting Practices, Third Edition
By Steve Conway

Logging Practices: Principles of Timber Harvesting Systems
By Steve Conway

The Logging Business Management Handbook
By Ronald R. Macklin

Directory of the Forest Products Industry

FOREST INDUSTRIES Mill Map

FOREST INDUSTRIES North American Factbook

FARMING THE SMALL FOREST

A Guide for the Landowner

Laurence C. Walker

A FOREST INDUSTRIES BOOK
MILLER FREEMAN publications

Miller Freeman Publications, Inc.

To
Janet Anne
Stephen Laurence
Wendy Elizabeth
Jean Marie
dear friends and good helpers, as well as the author's children

A FOREST INDUSTRIES BOOK

Library of Congress Catalog Card Number: 88-80192
International Standard Book Number: 0-87930-173-2

Printed in the United States of America.

CONTENTS

FOREWORD

America's increasing population, along with that of the rest of the world, calls for dramatically increasing rates of wood production if the nation and its export customers are to meet consumer demands for houses and paper. More and more will the country's policy-makers in government and industry depend upon non-industrial private forest (NIPF) landowners to supply the logs for these two principal uses of wood and for the hundreds of other products of lesser importance.

Commonly called "small timberland owners," this group of people includes farmers living on the land and a large array—lawyers, physicians, merchants and farmers—of absentee landowners. I have written this book expressly for them. I hope county agricultural agents, extension foresters in government and industry, consultants and others who advise the small landowner will find it a useful tool for better acquainting their clients with the universally vital renewable natural resource to which they hold title.

This writer, like every author, is indebted to many who, along the way, have—usually unknowingly—provided insights into the subject presented. Their number is legion. But I must here say thanks to Jamie Taylor, Stacy Shaw and Michele Allen for their diligence in transcribing into print the seemingly hieroglyphic handwriting of the author. I also express my gratitude to Dr. R. Montague Whiting, Dr. David Kulhavy, Billy Lemons, Gary Burns, Ernie Smith, Charles N. Spears, Ted Blackman, and to several senior classes at the School of Forestry, Stephen F. Austin State University, for critiquing portions or all of the manuscript. Jean Walker provided many drawings. Other illustrations are acknowledged in the captions.

Charles Taylor, editor of the Texas edition of the *Farmer-Stockman*, graciously gave permission to reprint with minor changes many articles which first appeared in that magazine. Some of Chapter 45 is reprinted from the author's "Ecologic Concepts in Forest Management," and some of Chapter 46 is lifted from his "Resource Managers and the Environmental Ethic," both first appearing in the *Journal of the American Scientific Affiliation*. Chemical Tables in chapters 17, 18, and 22 are from the *Forest Farmer 25th Manual Edition*, published by the Forest Farmer Association in 1985 (Chemicals used in Forestry Operations, by L. C. Walker, pages 40-43). The Glossary is taken from *Trees: An Introduction to Trees and Forest Ecology*, also by the author and published by Prentice-Hall Inc. in 1984.

The term "Tree Farm," occasionally alluded to in the text, and "The American Tree Farm System," are copyrighted by The American Forest Council. Beyond those incidental allusions and three photos courteously provided by the Council, no connection between this book and the Council is implied.

The author will appreciate receiving from readers comments which will make any future editions a more valuable tool for the use of nonindustrial growers of trees.

L.C.W.
Nacogdoches, Texas

C H A P T E R 1

TREES FOR TOMORROW

"Timber Famine" is a cry heard periodically throughout the land. In the 1870s, Congress reacted to the threat of depleted forests with the creation of the forestry division in the United States Department of Agriculture and the passage of the Timber Culture Act. The responsibility of the forestry division was to encourage landowners to manage their forests; the legislation assigned lands from the public domain to farmers and ranchers if they would plant trees on 40 acres of the 160-acre federal gift. The program failed because much of the land on which the "wildling" trees were planted was not originally forested, the species of trees were not specified, and nurseries to produce seedlings were lacking. But Americans were alerted. Conservation measures then began.

Again, timber famine threatened in the early days of this century. And again Congress responded by organizing the U.S. Forest Service and transferring vast blocks of federal forests from reserve status to that agency. Now, as national forests, they were to be managed in order to provide "for the greatest good of the greatest number in the long run."

Time was not long until the Weeks Act (1911), specifically designed to encourage the Forest Service to buy lands to protect navigable streams from silting in, also promoted timber production. A little later, in 1924, the Clark-McNary Law was passed, enabling the government to purchase private lands for inclusion in the national forests to assure a fiber supply for the future. By the last two acts and a few other Congressional measures, over 50 million acres were added to the national forests. Much of the new acreage, sometimes referred to as "the lands nobody wanted," was in the East and South.

In the West, the situation was different. The national forests remained somewhat like reserves, in spite of the transfer of the agency from the U.S. De-

Black walnut, the North American hardwood tree of high value for furniture, *U.S. Forest Service photo.*

partment of the Interior to the U.S. Department of Agriculture in 1905 and the subsequent change of the name from reserves to forests. Sustained-yield units were established, the first at Shelton, Washington, in 1946. These units, carved from national forests and in combination with private ownership, were to be used to guarantee a perpetual wood supply for a particular mill. The future would be assured. The mill would continue to run, laborers could be confident of employment, and communities would not die. As forest economics and forestry education gained an understanding, establishment of new units would not be required.

In the South, timber famine was again the headline in the 1930s as the last of the virgin forests fell to the steady cadence of the cross-cut saw, accompanied by the drumming of the double-bit axe. Cut-out-and-get-out was for real as sawmills ceased their whining in the Great Depression years. Mill closings came not so much due to the economy but, rather, because of lack of raw material to feed the carriages of the thousands of mills that dotted the Southland.

As with the earlier cries of timber famine, the nation responded as the last of the old-growth pines and hardwoods disappeared from the woods to reappear as studs and siding for homes across the land. Fire prevention, fire suppression, tree-planting by the "boys" of the Civilian Conservation Corps and the appearance of the Forest Service shield on boundary lines of thousands of acres of stripped land encouraged the regrowth of forest. Second-growth stands of timber brought to many locales a new interest in woodlands as economic enterprises.

Shortly, industries across the country assigned hundreds of privately employed foresters to encourage and assist landowners in proper forestry practices. Foresters employed by state forestry services and by agricultural extension services did likewise. The effort was expedited by President Eisenhower's soil bank appropriations in the 1950s, when the government paid farmers to replace cotton patches and food crops with pine trees. For the soil bank program, farm stability, not wood production, was the incentive.

That threat of timber famine, like those of earlier times, was over for a while. The threats have been called a hoax, a way to get more money from government for forestry, and a cause for anger by those who in the 1960s were unable to get a good price for their stumpage (the price of trees standing in the woods).

It was this glut in the marketplace—cheap wood—that encouraged industrial consumption to feast on the supply. New mills were built, old ones expanded, and new products developed which could be manufactured from forest fiber, especially from the trees of the southern region.

The bounty was to be short-term. Now, in the late 1980s, timber famine is a threat greater than those warnings of the past. America's ever-increasing population, along with the demand for more products from trees by each of the many more people, has seriously eroded the timber base. The number of Americans we officially count increases by 2% per year (this number excludes the illegal immigrants). Wood consumption rises by 3%. The high capital investment in equipment today compared with the "peckerwood" mills of a generation ago adds to the seriousness of the matter. The saws must "run wide open" to pay for the investment.

This time, the timber industry and government attack timber famine in a different way. Both offer financial incentives. Companies in one state banded together to provide money for small landowners to prepare sites for planting, to plant trees and to kill weed trees (see Chapter 3). The federal government has FIP, the Forestry Incentives Program (see Chapter 2).

Some forestry leaders consider the matter so serious they want legislation which would require, under the penalty of a fine, regeneration of cutover lands within two years of harvest. Currently non-industrial landowners in the South plant perhaps 1% of the land they harvest each year. Elsewhere, it is little different.

Five men appointed by President Nixon called for a national forest policy council. They also pointed out the need for governmental assistance to the owner of small woodland acreages, usually the farmer, but often the city-dwelling absentee possessor of rural acreage.

The report carried this quotation:

> And this devastation (of our forests) is now become so epidemical that unless some favourable expedient offer itself, and a way be seriously and speedily resolved, for a future store, one of the most glorious and considerable bulwarks of this nation will, within a short time, be totally wanting to it.

These words, however, did not originate in the 1973 presidential panel's report. John Evelyn wrote them in 1662. While Evelyn wrote his *Silva* for the British Royal Society, others expressed similar thoughts about the forests of the New World. An early 1800 report noted, "The most urgent motives call imperiously upon (the U.S.) Government to provide a reasonable remedy for the evil (of short timber supplies)."

The President's panel, convened in the fall of 1971, was to study the entire range of management problems on America's forests in order to advise the President on how to meet the growing housing needs while "protecting and enhancing the quality of our environment."

Congress continued to recognize that caring for the nation's privately owned forest lands is in the public interest and, therefore, is a proper use of tax funds. Many acts authorized expenditures of federal money for protecting woodlands from fire, insects and disease. Cost-sharing, too, for improving the condition of timber stands was agreed upon. President Nixon's advisors recommended continued financial support for these purposes "at appropriate levels."

The controversial capital gains tax was encouraged. Tree farmers until the end of 1986 did not treat income from timber sales as ordinary income, which is taxed at a higher rate. The panel recommended, without much elaboration, other "special" tax provisions to further encourage good forest land management for non-industrial owners. Long growing periods, high risks, and modest rates of return on forestry investments make this desirable.

"If public programs to increase wood growth . . . are undertaken, they should be concentrated on softwood forests." This means, the panel says, "pine forests in the South." Construction lumber and dimension timbers, industrial plywood, and paper pulp are almost wholly manufactured from conifers.

Currently about 3.5 billion board feet (bd ft) are harvested each year. Increased production will come from thinnings, sanitation cuttings and the salvage of dead and dying old-growth, measures designed to enhance the remaining timber. The panel members insisted on bringing about full-stocking immediately on government forests.

The report of the Nixon-era panel was severely criticized in the press as being "pro-industry." Much of this, strangely, emanated from the western timber-producing states by newspapers dependent on wood fiber. Some editorial critiques jumped over the broad implications of the policies suggested to harp upon a minor point—such as scenery—which became fashionable to favor. Panelists emphasized that this was not a report on "multiple-use" management; rather it was about how we get timber for America's future.

One hopes a president will see fit to create a permanent forest policy council that will be receptive to the needs of all Americans in pursuing proper management methods for public forests, industrial ownerships and small private woodlands. President Nixon's statement upon receiving the document was symbolic of continuing national concern. He concluded, "All Americans should recognize the value inherent in the unique renewability of our forest resources; they must be cherished, nurtured, and intelligently utilized. This report is a valuable contribution toward helping us meet these goals."

A few statistics may aid the reader to appreciate the national significance of wood production on non-industrial privately owned forestlands (NIPFs). For owners of these tracts, this book is written.

NIPFs in the U.S. comprise 300 million acres, or 58% of the total forested area. Almost 8 million ownerships, two-thirds of which are 10 acres or less, are involved. These 10-acre-or-less parcels, however, account for only 4% of NIPF acreage. Only 1% of the owners hold 500 acres or more, but their holdings amount to 34% of the NIPF total. Non-forest-industry corporations, youth groups, and recreation clubs (all considered NIPF) control over 40 million acres. Two-thirds of the nation's forestland is owned by farmers, laborers, business and professional people and retirees. Government and the wood-using industry claim the balance.

In Georgia, less than 15% of the timber cut in 1983 was from NIPFs, although the 11,000 owners held almost 4 million acres. In West Virginia, retirees make up 25% of NIPF owners. In Florida, NIPFs amount to more than 50% of the commercial forest acreage. New England's five states have over 500,000 owners eligible for federal assistance programs.■

CHAPTER 2

FEDERAL INCENTIVE PROGRAMS

Small landowners now have an opportunity to make significant contributions to the nation's wood basket with help from the U.S. Treasury. To encourage the production of wood fiber for the future, financial assistance for forestry practices, funded from federal taxes, is available to people owning less than 1,000 acres of land in any one county capable of producing commercial conifers . Up to $10,000 may be allocated to a single owner to pay 50% to 75% of the costs of several approved practices. As details change with the years, landowners should obtain current requirements from County Agricultural Stabilization and Conservation Service (ASCS) offices.

Practices encouraged by this money include site preparation—preparing the land for planting, tree planting and conversion of low-grade weed species (such as scrub oaks) to high-quality stands (such as pines). Direct seeding on clearcut or burned-over land and the use of chemical injectors to reduce the number of undesirable stems in mixed-species forests are other appropriate practices. Prescribed burning for certain purposes is also authorized. FIP, for Forestry Incentive Program, makes money available to the landowner—depending on desired practice, the state in which the land is located, and management prescriptions. The amounts range from $20 to $150/acre.

The FIP program, initiated in the latter Nixon years to assure American independence for wood in the year 2000 and beyond, was originally allocated to aid owners of productive woodlands of 500 acres or less to plant trees and to control weed trees. Because much of the congressional appropriation assigned to some states in some years was not used, interest in the program did not seem as strong as anticipated. Rules for managing FIP were liberalized for 1978 so that state foresters and state ASCS

A mixture of perhaps 50 hardwood species encroach in a southern pine stand. The federal Forestry Incentives Program (FIP) provides funds for deadening these weed trees, *U.S. Forest Service photo.*

committees may, by special request and approval, permit farms and ranches with up to 1,000 acres of woods to participate. Almost half of the 500 million acres of the nation's timberlands are in ownerships of this size.

Generally, the waiver of the 1,000-acre restriction is limited to association, group, partnership, and undivided estate ownerships. But individuals who own between 1,000 and 5,000 acres and who will plant at least 40 acres of trees or do timber stand improvement (TSI) work on 50 or more acres are eligible for waiver consideration. About 200,000 acres are reforested or receive some sort of weed-tree control each year.

Weed-tree control, the principal TSI practice, may involve chemical herbicide applications, mechanical site preparation or prescribed burning. Reforestation may be by tree-planting or direct-seeding. Planting hardwood (broadleaf deciduous) trees is no longer an eligible practice in some areas. Other changes in the rules include establishing a minimum of 10 acres of woodland ownership for eligibility and the same acreage as the minimum area for reforestation.

Although cost-sharing for repeated prescribed burning on the same site is not permitted, owners may apply for funds for site preparation and for subsequent planting on the same site. This makes good sense: land left idle after clearing is wasted and quickly captured by both herbacious weeds and brush.

Under this "two-component" practice, the landowner may choose to control low-lying brush with chemical sprays and return a year later to the same tract to kill larger worthless hardwoods with herbicide-loaded tree injectors. Both tasks could be funded; and, if fully so, only one of four dollars comes from the landowner's purse.

Practices specifically prohibited by the rules are the planting of Christmas trees, as well as hardwood species. (Shelterbelts are of course excluded, for the program is operating only in counties with commercial forests and not in the plains.) Otherwise, all species available in state-operated forest-tree nurseries meet FIP stipulations.

Fencing is not a cost-share activity, but could be a required item. The barriers can be permanent or temporary, aimed at protecting the trees from hazards such as wildlife or livestock browsing or trampling. However, as most trees in most forests are injured by cattle, sheep, hogs, deer and rabbits for only short periods—during the seedling and sapling stages—permanent fences are unnecessary. Since the rule-writers acknowledge this, it is unlikely fencing would be approved. If it should in the future be acceptable, the federal part would be limited to cases of especially high-value stands. (Wild black cherry and black walnut would have been good examples when hardwood plantings were subject to approval. Because of the scarcity of these prized furniture woods, cost-sharing for silvicultural procedures and fencing could again become appropriate.)

Precommercial thinning is eligible under FIP. Thinning that results in the landowner receiving money for wood sold from the tract is not.

That is unfortunate, for commercial thinning is a much-needed practice that should be approved for cost-sharing. Great quantities of wood are wasted because of failure to thin forests, especially in zones with fast tree-growing climate and soil. Trees that should be removed for pulpwood, fence posts, chips, or even sawlogs die "on the stump," eventually being thinned by nature. That is waste.

Because these stems have not been removed, trees that would have been residuals are not released from competition for soil moisture, nutrients and sunlight. These better trees, ordinarily designated for later harvests, are thus not making maximum growth. Hence, the wood they are *not* growing is also a waste, in this case the waste of the site.

Late evidence strongly suggests that outbreaks of the southern pine beetle and other insects are related to high stand density. Abundant food, providing a favorable habitat, becomes available to insects if forests are not thinned. If infested stands had been thinned, this epidemic, killing millions of board feet of good timber in the South, could have been averted. Perhaps outbreaks of the destructive mountain and western pine beetle infestations in the forests of the Rocky Mountains and the Far West may also relate to a lack of thinning.

Stands are not thinned simply for economic reasons. Loggers generally can remove a unit of logs or pulpwood so much cheaper in clearcutting operations than in thinnings. A proposed FIP provision, perhaps subsidizing the logger, would help to remedy this situation and thus more fully assure for another generation adequate timber supplies from better-managed forests.

FIP has had direct and favorable effects upon another fairly new industry: consulting forestry (see Chapter 35). Practicing consultants receive numerous referrals from state foresters who find that a woodland to be examined exceeds, for one reason or another, their authorization to carry out the task. Both the state employees and the consultants write management plans and guide, or perform, planting and TSI work.

Especially important to landowners with inadequate capital to hire forestry consultants is the restriction of how much time foresters of a state or federal agency may spend with an individual on his land. The FIP rule governing this aid is to "improve utilization of the limited amount of technical assistance." To facilitate availability of professional expertise, branch offices with state-employed foresters and technicians in residence have been opened in many counties of the forested parts of the country.

How does the landowner take advantage of the FIP opportunity? First step is to enroll the land and apply for the practice at the county ASCS office in which the land is located. There one will learn the cost-share limits, based upon forester estimates, and the amount established by the state and local committees.

A forester, privately employed or with a state forest service, is then asked to visit the land and prepare a management plan. That plan, approved by the state forest service, is sent to the ASCS office for processing and the earmarking of funds for the requested tasks. Payment is made upon completion of the work and approval by the state agency.

Industrial foresters may also provide assistance and make management plans for landowners interested in federal programs. Conservatively, at least one million timberland owners in the United States are eligible to participate in this program. They own about 300 million acres of commercial forestlands. The industrially supported Tree Farm System includes 40,000 members, owning 80 million acres.

The State Forester designates the counties eligible to participate and recommends how much money should be made available to each county. ASCS is charged with the responsibility to assure that these funds are not transferred to other farm uses and to make certain that FIP funds are not allocated above the maximum of $10,000 for a single ownership. In contrast to other farm programs, FIP legislation and the record of congressional hearings make it clear that this is not a social program in the ordinary sense. Monies are not to be prorated or manipulated in an attempt to get more pieces out of the pie.

Currently, ASCS signs up landowners on a first-come, first-served basis in order to dispense effectively annual allotments by the end of the federal fiscal year. The money need not be spent in the year allocated: it must, however, be earmarked for a particular task.

Industrial landowners are not eligible; but groups or associations, such as hunting clubs, are eligible if they do not own more than 5,000 acres of qualified woodlands. This is a small landowner program, but the landowner may be absentee.

An important consideration to note is that, unlike the REAP (Rural Environmental Agricultural Practices) Act program, also handled by ASCS, FIP decisions need not relate primarily to soil and water conservation. To enhance yields of clean, clear water, or to retard erosion is important, but not the first priority in the landowner's decision-making for FIP. Increased production of wood fiber for the future is the only goal.

Hence, the federal money cannot be used to establish a private wilderness. Present law says the land must remain in timber for at least 10 years. Non-commercial cutting at any time, such as a thinning to release the best trees, would be considered appropriate.

Patience may be a needed virtue. Now that the program is running, many cases are pending which, if completed, would utilize the funds sent to the states. Experience suggests, however, that many projects will not be completed. The discovery by landowners, for instance, that the cost-share grant is taxable as ordinary income discourages some from following through. The 1986 "simplification" federal tax act did not change this. (As I write, a ruling by the Secretary of Agriculture leaves this matter in limbo. These payments may, or may not, be taxed as ordinary income. Tax courts will have the final word.)

FIP work on the small forest is not totally federally financed. ASCS may provide up to 75% of the cost of the work, but is not obligated to do so. How much money is available, how many landowners are making requests, and how many acres are likely to be involved are determining factors.

Estimates indicate that two-thirds of the landowners are "willing" to invest some money in forest management; but only 10% are likely to do it, and then only on a cost-share basis with government. Those participating in such a program may be interested in earning personal income from later timber harvests, either for themselves or their heirs. Some landowners are likely to have more altruistic motives, such as soil conservation or wildlife preservation. In either event, the purposes, as established by Congress, will be well-served if the nation's wood box is better filled when today's youngsters reach three score and ten.

Costs for these jobs to the landowner have been averaging about $40/acre, though varying from less than $25 to more than $200. The federal cost-share average exceeds $50/acre.

What of the future of FIP? A two-week annual sign-up period with the ASCS is proposed, but likely to be resisted by state foresters as tidy paper work and unrealistic if the goal is to serve the public. A major factor in Washington balloting continues to be how efficiently the funds are used.

The loss of the capital-gains tax break occasioned by passage of the 1986 tax reform act is expected to affect the willingness of non-industrial forest landowners to invest in raising productivity. To compensate, Congress will be encouraged to provide other incentives. Among these could be a stronger Rural Forestry Assistance Program (RFAP) and Agricultural Conservation Program (ACP). RFAP enables state and federal cooperation for multiple-use management, including watershed management and recreation. ACP permits federal cost-sharing for programs beyond FIP.

Another area for funding is through the Agricultural Conservation Program (ACP), also handled by the U.S. Department of Agriculture's ASCS. With ACP, no acreage limit applies, although payment to any one person in a fiscal year is limited to $3,500.00. In contrast to FIP, for which the aim is wood fiber production, ACP is designed to encourage placing agricultural land in a conservation reserve. The landowner's intent is the qualifying factor as to which program to choose.■

C H A P T E R 3

STATE AND INDUSTRY INCENTIVE PROGRAMS

Motivating owners of small forest tracts has concerned American foresters and the wood-using industry since the beginning of the conservation movement. In tree-growing regions, such owners rarely invest in forest conservation measures. As a result, their woodlands grow less than half as much wood per acre as industrial forests. Since farmers, ranchers, and other non-industry owners control over half of the nation's commercial woodlands, production of timber on their lands must rise if the nation is to meet its wood-fiber requirements a generation hence.

As the nation grew, so did the consumption of wood products. In the 1930s, a Senate document, called the Copeland Report, told of threatened timber shortages. Greater efforts to manage our forests followed. Some money was earmarked to encourage forestry on private lands.

In every period of threatened timber famine, the idea of stimulating farmers and other non-industrial landowners to grow trees was apparent. Beginning in the 1930s, too, the Forest Service and the Soil Conservation Service established demonstration farm forests to show owners of small tracts that forestry could be profitable. Unfortunately, at that time, few landowners considered the economic benefits attractive. Nor are they likely to do so today without a financial incentive.

Congress, encouraged by the paper and lumber industries and the forestry profession, hopefully found a way. The Forestry Incentives Program (FIP) (see Chapter 2) and tax breaks (see Chapter 29) came into being. Industries looked for alternative programs.

Industry in Texas provided an alternative. So has the government of Virginia. The procedures are similar, both modeled after FIP. By 1990, extension of both the state and industry programs are expected to be common.

Industrial grants are matching-money programs, the owner paying 25% of the cost of the approved practice. Usually two strings accompany the allocation: minimum tract size of 10 acres and the retention of the effort (such as the trees planted) for at least 10 years. Those who manufacture wood products are not eligible to participate. The maximum amount available to a landowner requesting a state's grant is $160/acre. Funds are available (a) to clear land as a site-preparation practice by prescribed fire or mechanically with light or heavy equipment, (b) for weed-tree control, (c) for pre-

State and industry incentive programs assist in planting pines on cutover and burned over lands belonging to small private forest landowners, *Langdale Co. photo.*

commercial thinning and (d) for planting.

The procedure is simple: any forester looks the land over and prepares a brief one-page management plan. This is approved by a state forester who routes the request for help to an appropriate committee. The committee, utilizing funds contributed by industry or the government, reviews the request and allocates the money.

In Texas, the industrial fund receives from a dozen principal companies about $1 million a year for distribution to non-industrial landowners. Applicants may contract with consultants (see Chapter 35) or do the work themselves.

Why does industry have this interest? They must have wood from which to produce wood and paper products for the nation's citizens. Only a few companies grow all their needed wood on their own holdings. Industry obtains as much as 50% of its raw material from small landowners. They may be all manner of people: farmers, ranchers, and business and professional people living on, or far from, the land.

What restrictions are placed on the landowner? He is obliged to protect planted and released seedlings from destructive fire and grazing. The earliest harvest time will vary by county; the life-span of each practice is a decision of a local committee. Until the final harvest, averaging perhaps 35 years after planting, the land for which the owner enjoys the financial incentive must produce wood for America's consumption.

Applicants for these industry and state grants should check with the state forestry service, whose foresters recommend silvicultural treatments to increase timber production on the particular parcel of land. The management plan follows a simple outline. It is free if prepared by the state. For expensing costs in federal tax returns, it is important that landowners make the decisions as to what shall be requested and done to the forest. Agency foresters approve the practice for cost-sharing. However, to speed application processing, landowners may elect to have consulting foresters handle this part of the application (see Chapter 35). (Texas does not permit consultants to complete its industry-assistance applications; State Service foresters must do this.)

Recommendations of the forester depend on the productivity of the land. High site-quality parcels get priority on funds available for a specific county. Sites capable of growing trees 80 ft tall in 50 years (Site Index 80) are virtually assured of receiving cost-share assistance. Lands where trees may grow to a height of just 50 ft in 50 years (Site Index 50) would not likely qualify for assistance if better sites in the county are available (see Chapter 26).

Management Assistance Programs (MAPs) and Land-owner Assistance Programs (LAPs) encourage nonindustrial private forest (NIPF) landowners to sign agreements with industry for the oversight of their holdings. Some company contracts require landowners to sell to the company their stumpage when it is ready to harvest. Other companies utilize leasing arrangements for NIPF land. Still others are satisfied with nonbinding agreements.

Leasing of NIPF lands appears to be on the wane because (1) owner objectives change during the long time period until trees are ready to harvest, (2) economic conditions are usually uncertain over the time-span, (3) contract payments do not keep up with stumpage and land values, (4) capital gains tax laws change and (5) nonbinding agreements provide the company about the same amount of wood at a competitive price without the strings attached to a contract.

Companies may lose out with leases because courts have allowed heirs to break the contracts and because disputes occur with the Internal Revenue Service over expensing. Contracts are not as effective as simple agreements for industry public relations, and the company can often get wood from other owners at a lower cost than the contracted price. Company personnel sometimes seem not to realize that lessees and contractees expect a profit amounting to more than ordinary compounded interest rates on their investments.■

C H A P T E R 4

NATIONAL FORESTS— THEIR MEANING TO THE PRIVATE LANDOWNER

About 186 million acres of federal lands are tucked away in national forests. Most of those tracts in the West were transferred from the public domain at about the turn of the century or from the old national forest reserves in 1905. Those in the East and South were bought, for the most part, from industries or farmers. At the time of the purchase of the latter parcels in the mid-1930s—for as little as $2 or $3 an acre—these tracts had little significance to the economy of the local area or of the larger region.

But that is no longer the case. What makes these federal "lands that nobody wanted" so important to today's citizens—whether one lives in the cities, owns a tree farm, or farms the High Plains—is a 1970s' decision of a district court in West Virginia which was upheld by a circuit court in Richmond, Virginia. The courts' actions, now popularly known as the Monongahela decision for the name of the national forest in those West Virginia mountains, ruled that the Forest Service violated its charter by permitting clearcutting. Indeed, the Act of June 4, 1897, as freshman forestry students would know, provided funds and directions for managing what were then the national forest reserves.

Those directives of Congress stipulated that every tree to be sold and harvested on reserve lands had to be large-growth, mature or dead. And each tree had to be individually marked, and every marked tree had to be harvested. No way could every tree in a clearcut zone be economically marked; and many small trees, culls, and those of unmarketable species must be left as commercially unmerchantable. Nor does a clearcutting regeneration harvest distinguish between maturity or "large growth" among trees. All timber that may be economically taken to the mill is removed from the forest, regardless of the size or age of the tree.

Forest Service officers, however, believed that later laws, including annual congressional appropriation bills, permitted the harvesting of all trees in designated blocks so that new forests of high-quality seedlings, those that would grow into commercially valuable wood, could be established on the cleared sites. The jurists did not agree.

The affirmation of the circuit court literally shut down timber harvesting on the national forests in the four states of its jurisdiction. Subsequently, the district court in Alaska enjoined the Forest Service

A dense stand of lodgepole pines in a national forest in the Colorado Rocky Mountains (U.S. Forest Service photo).

from continuing its 50-year contract to supply wood for certain industries in that state and in Japan.

A similar contest in Texas could have been catastrophic to both small and large industries—the two-man independent logging crew as well as the wood-using giants—had the decision of the district court in the Lone Star State not been overturned by a higher federal appeals bench. It could also have pinched the purse of all Americans.

About 20% of the wood used in America's pulp, paper, plywood, lumber and pole plants is grown on federal land. These forests include tracts managed by the Bureau of Land Management in the U.S. Department of the Interior as well as national forests, lands administered by the Forest Service, a bureau of the Department of Agriculture. That harvest amounts to approximately 20.3 billion bd ft a year.

If this wood were not available for a year or two, the injury would be limited to those companies that do not have ready access to other supplies and which depend upon federal timber for survival. A long-term moratorium would be felt by all buyers of timber, resulting in higher prices paid in a more competitive market.

Longer hauls from available forests to mills formerly supplied from federal lands would also cause higher prices for lumber and paper, since transportation from woods to mill is an important ingredient in the recipe for calculating the cost of wood to the consumer. This, of course, would mean raising the price of the 2x4 stud and the cost of a house as well as the cost of this book.

Currently, about 16% of the nation's timber is cut from the 156 national forests in 44 states. Federal forestry officials predict that had the West Virginia Circuit Court's decision been applied to the whole country, national forest harvests would have been immediately reduced by 75%. After a year or so, for the rest of the century, the cut would be half of that now removed each year.

Most of these government lands are in the West—from the Rocky Mountains to the Pacific. There, in contrast to the southern coastal plain region, trees grow slowly. Several hundred years may be required for some species to reach maturity on some sites; while in the South, around a hundred years is considered about maximum. Under management and on favorable sites, economic maturity may be obtained as early as age 40. In the South, age 30 is not too unusual.

As the time for harvesting trees in the West is extended to comply with clearcutting restrictions, and as a far greater percentage of forest land is in national forests in the West than in the East and South, much more will be the demand for non-government timber throughout the country. That demand must be supplied by wood imports or by domestic private forests.

The South is now developing its "Third Forest" of planted genetically superior trees which will be carefully managed for perhaps 30-year rotations. The region, therefore, is not prepared to provide on short notice the needs for fiber which, until now, have been taken from western national forests. If it comes to that, southerners should expect serious overcutting for a short time to accommodate citizen needs for paper and housing. Meanwhile, western corporate owners, to satisfy a potential shortage, are growing and planting seedlings, intensively controlling forest weed trees and brush, regenerating ragged old stands and fertilizing forests.

There is, of course, an alternative: follow the courts' mandates to abide literally by an obsolete law, drafted before the U.S. had even one professional forestry school and but a few native-born foresters. The two courts probably little realized that few of the forest trees on managed lands in the East or South are large-growth or mature. If they grow old or mature, we have been poor managers. To be economically useful, they must be harvested before they reach maturity. The day after maturity, they could be dead; dead trees in the warm, moist southern forests decay too quickly to be merchantable. To harvest only large or mature tomatoes or hens may exaggerate the idea, but it is the idea.

Now, if every tree in an area were large or mature, to mark them individually would be costly; and this expense must be borne by the ultimate consumer. Thinnings, too, are prohibited by the law as interpreted. It is the thinnings which provide some interim income before final harvest, while releasing the best trees to approach maturity quickly.

There are other problems. Foresters have no yield data for forests managed according to the 1897 law, never dreaming that a court, and a Congress long reluctant to act, would in effect toss out the millions of dollars spent in research and the experiences of 80 years of practice. No one knows how much wood our forest lands could produce under those conditions.

Administrative expenses for the Forest Service to oversee partial harvests would be almost doubled. The management costs of timber sales would increase by 25%. Finally, no satisfactory conclusion has ever been reached as to the intent of the word "mature" in the 1897 law. Did the legislators mean financial maturity, physiological maturity, age maturity, or sexual maturity? For a particular tree, sexual maturity—when the tree produces fruit and seeds—may be at 10 years of age, economic maturity at 50, age (rotation) maturity at 60, and physiological maturity at 70 years.

The 1897 "organic" act, passed before there were any national forests and aimed at virgin timber stands of the West, was intended to protect the forest reserves. Only later, beginning about 1907, did the idea take hold to use these lands for "the greatest good of the greatest number in the long run."

This greatest good was generally thought to be timber production, extending through periods of wood scarcity and adequacy. Now, however, multiple-use, a concept approved by Congress, requires that management also consider recreation, wildlife, water and range when making plans for the use of national forest land.

Because these forests comprise over half of the country's commercial softwood timber holdings, it is doubtful if Congress will actually provide the money to give each use equal measure. Timber will continue to be "first." Indeed, sales of timber now bring into the federal Treasury about 80% of the Forest Service timber management budget. Congress may talk multiple-use, but the non-timber components will remain secondary because money often speaks louder than Congress.

National forests are important in another way. One-fourth of the income from the sales of timber (or grazing leases or recreational use permits) is returned to the counties from which the wood was harvested. In some years this amounts to well over $1.5 billion. The return to counties in the especially-low harvest year of 1986 was about $270 million. Sometimes called "in lieu of tax" money, as the U.S. Government pays no local land taxes, it is earmarked for schools and roads. State legislatures divide the money between the two uses. Another 10% remains locally available for road improvement within the forests, while a variable amount is placed in a trust fund for improving timber stands. All of this designated money puts local people to work.

While sales of wood from national forests tend to hold down the prices of goods made from wood, they also hold down the stumpage price which the nonindustrial private forest landowner may obtain for his trees. As management costs for the national forests are subsidized from the federal treasury, taxpaying owners of small forested tracts help provide funding for a giant competitor in the marketplace.■

SILVICULTURE

From the Latin, *silva*, trees; and *cultura*, art; thus, "the art of growing trees in managed stands for the production of wood, water, wildlife, and nonmaterial amenities"

C H A P T E R 5

RENEWING THE FOREST— CLEARCUTTING HARVEST SYSTEM

The cutting of trees in a managed forest at the end of the rotation is sometimes called the "final" harvest. It is not, however, the end; rather, it is to begin a new forest that trees are cut and sent to the mill.

When trees reach maturity, they should be cut to provide wood for man's needs and money for the owners. Trees may reach economic (sometimes called financial) maturity long before they are old or cease to grow. Economic maturity is the time when the landowner can earn more by cutting his trees and putting the money received in the bank to draw interest than he can by leaving them in the woods to grow.

Trees, in other words, are the capital investment; the wood they add each year is the interest on that investment. At some point, for every tree (or every stand of trees), the interest rate in relation to the capital declines to a point where failure to harvest is wasteful, both to the landowner and to society. Followers of the stock market call this the price:earnings ratio. The price in the case of a tree (or the stand) is the value of the tree (or the stand) growing in the woods (stumpage); the earnings is the value of the wood laid down each year in height and diameter growth in the bole (or for all the trees in the stand). The farmer may think of over-mature trees as he would over-ripe corn. The banker may think of them as an over-mature, or past-due, note.

Physiological maturity, on the other hand, occurs when, for all practical purposes, the trees cease to grow. It could be decades after they reach economic maturity. Such trees may be pleasing to the eye and produce food for wildlife, but they no longer add an appreciable amount of wood fiber to the bole. Likely to be especially susceptible to insect and disease attack, their retention may result in less wood in the inventory of the forest. Often rot in the interior of the stems is hidden from view.

Our concern in this chapter is clearcutting. As the name implies, all the trees on a parcel of land are harvested at one time. For the small forest, this could be in blocks of 10 acres, seldom more than 40 for natural regeneration. Size of the opening will depend upon the species, especially how well its seeds travel. Heavy-seeded trees, like the hickories, hardly can be expected to reproduce in openings larger than a portion of an acre; while the lightest seeds among the winged pines and maples may be carried by the wind considerable distances to regenerate a large tract. With clearcutting, seeds come from the walls of trees surrounding the opening.

Ordinarily, a clearcutting regeneration harvest area should be separated from other areas that are

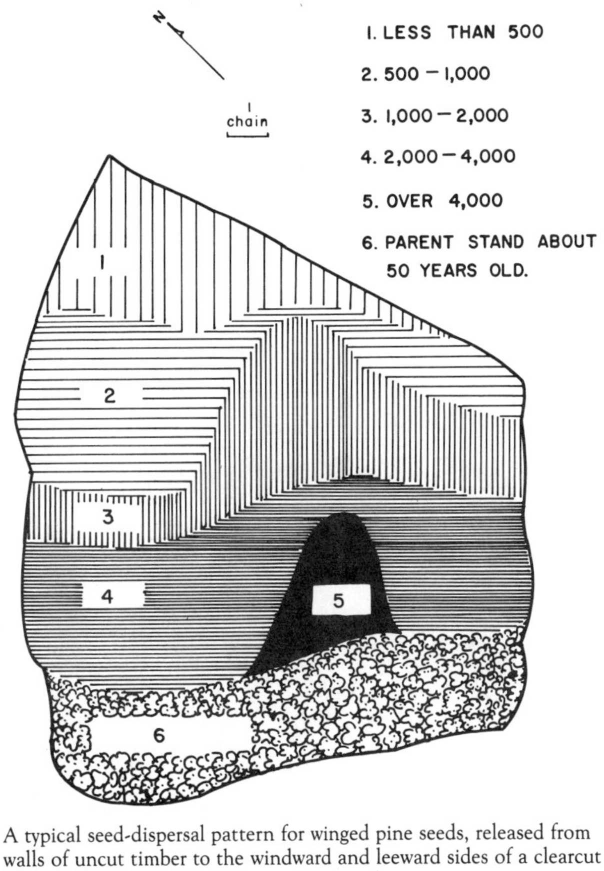

A typical seed-dispersal pattern for winged pine seeds, released from walls of uncut timber to the windward and leeward sides of a clearcut woodland. Here the wind is out of the north.

to be harvested the same year by distances of at least three times the width of the proposed clearcut opening. Thus, if a landowner's 100-acre tract is to be clearcut in 10-acre patches, perhaps three such openings could be spaced throughout the tract.

Clearcutting may be in strips as well as patches or blocks. If in strips, these should be oriented in the direction of the prevailing wind so that the opening serves as a corridor for the wind. A strip laid out to block the wind, like a windbreak, may suffer from wind-throw. Yet clearcut strips may be laid out perpendicular to the wind's direction. In that event, the first strip harvested is on the lee (downwind) side of the tract, subsequent cuttings made further away from the lee, but with a residual forested strip between.

The walls of timber surrounding a clearcut lend themselves to wind-throw. Thus, trees with shallow root systems growing on peaty soils and trees without tap roots growing in shallow mineral soils may suffer severely from wind damage.

Clearcutting is ugly; there is no denying that. Until new seedlings are established, the site may look like it has been blitzed. It is, however, a legitimate regeneration harvest system, introduced to European forestry by German foresters several centuries past. Clearcutting should not be confused with the rape of the land, the indiscriminant cutting of all of the trees without any thought for establishing a new forest. State and local laws may soon prevent clearcutting because of the environmental insult that follows, even if for but a short term.

Landowners often believe that tree planting must follow clearcutting of a forest. This expensive practice need not be necessary. Carefully timed harvests—removal of the old forest and scarifying the soil in the process just before seeds fall—provide for a new natural-appearing forest.

Direct-seeding, usually much cheaper than planting, also allows for renewing of a woodland that appears natural, in contrast to the orderly rows of trees in planted forests. Planting and/or direct-seeding would be required if the purpose of clearcutting is to establish a new forest of another species or one of a higher genetic quality.

Clearcutting results in even-aged stands. Some species, like the southern pines and densely growing ponderosa pine, are typically of a single age within a stand, even though the trees in the stand may not look the same age. Such an appearance occurs because trees growing side by side may vary in diameter from 3 in. to 30 in. In all likelihood, if the species is intolerant of shade, they seeded-in from the same seed crop.

While clearcutting is always designed to reproduce a forest, its timing may depend upon circumstances beyond the landowner's control. Serious outbreaks of insect infestations (like the periodic

A clearcut coastal forest, later drained to encourage reproduction.

outbreaks of pine beetles in the West and South in recent years), disease infection (blister rust on eastern and western white pines and annosus root rot on slash pine in the South's coastal plains and Douglas-fir in California), or physiologically overmature trees (the old-growth, virgin forests) may dictate the necessity to use this method. Salvage of trees killed by fires, floods or other natural catastrophies requires clearcutting.

Prescribed burning may be necessary to scarify the soil to receive seed and to eliminate logging slash. Untreated slash in the West often is a considerable fire hazard for many years for the newly established seedlings. In the South, logging slash rots away within a few years.

If blocks rather than strips are clearcut, an effort should be made to have their edges follow contours. This gives an uneven, more natural, appearance to both the harvested and residual areas, in contrast to the squared, precise look of blocks laid out on surveyors' section lines.

Landowners often prefer clearcutting because it may be the least expensive method for the logger and, therefore, provides the owner with the highest income for the timber removed. Loggers with minimal skill are able to handle the task because there is no risk of damaging residual trees. Also, carrying capacity for grazing animals and wildlife habitat may be enhanced.

Among the distinct disadvantages not already alluded to is the possibility of increased soil erosion and the consequent siltation of streams and lakes. Clearcutting may detrimentally affect water quality and fisheries. If natural regeneration is depended upon following clearcutting, seedlings will not be as evenly spaced as in planted stands. They may be bunched in pockets due to especially desirable or undesirable micro-environmental conditions and because of the manner of seed dissemination. The temporary loss of seeds to sustain wildlife also may be a disadvantage in using the system.■

C H A P T E R 6

RENEWING THE FOREST— SEED-TREE HARVEST SYSTEM

Not quite so devastating in its appearance as clear-cutting is the seed-tree regeneration harvest. Leaving five to 10 trees well-spaced on each acre provides the "bull seed" for the new forest.

As these seed trees will be the parents of the next stand of timber, they should be selected for characteristics other than their location among their neighbors in the particular forest to be regenerated. Fast growth and good form likely first attract the observer. Because seed-tree harvests generally take place in even-aged stands, the larger trees are obviously those that have grown the most rapidly. Trees 3 in. and 30 in. d.b.h. (diameter, breast-high {4½ ft above the average ground level}), side by side, may well be the same age. Form deals with straightness and taper. Trees with crook, often an inherited characteristic, and with much taper—more like a candle than a rolling pin—may pass those traits to their offspring.

Some species inherit resistance to insect and disease attack, so stems evidencing such maladies rightly may be passed over as seed trees. Douglas-fir, which is often infected with a rot, however, does not fit this pattern. For that species, stems with the disease should be left in the forest to produce seeds. Such trees at the mill may have little value, the bole being filled with rot.

Seed trees must produce seed. For conifers, this capability can sometimes be determined by cones persisting upon trees in the woods. Other times, for species without persistent cones, one may find them on the ground beneath the crowns of prolific producers of fruit. Often all that is needed to get trees to flower is to remove their competition for sunlight and shade. Choosing and releasing from competition the seed trees may enable adequate seed production a year or two after the harvest. If there is no evidence that any of the trees in the woods have abundant seed-producing capability, the landowner may need to consider fertilizing the chosen seed trees with nitrogen. Applications of about 5 lb of ammonium nitrate (NH_4NO_3) under the crowns and a little beyond of such trees may bring about flowering of buds. Fruit and seed production should follow.

Major problems occur for stands treated by this method. The most serious is lightning strikes that kill the lone trees silhouetted against the sky. In parts of the country where dry lightning storms occur, fire may spread through the logging slash left from the harvesting operation and kill reproduction as well as the tall, valuable trees.

Another problem is the need to remove brush

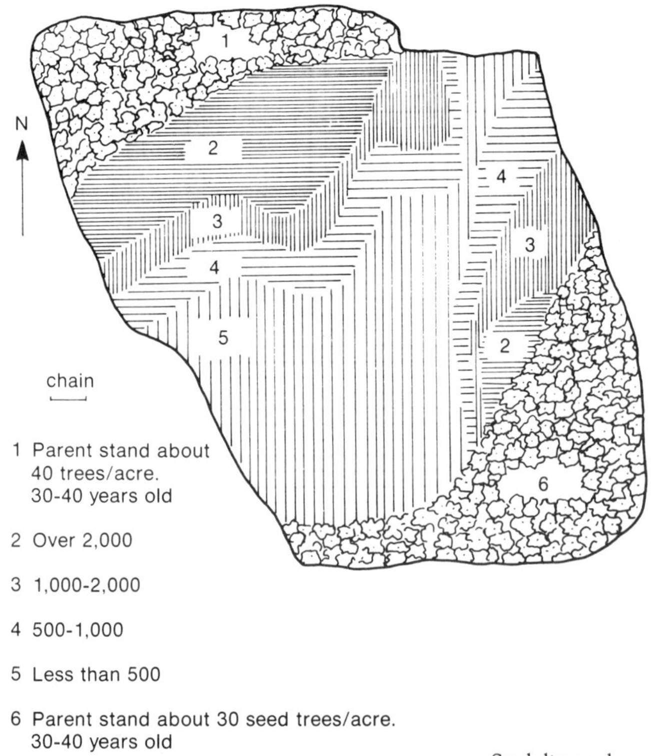

Seed dispersal after a seed-tree harvest.

where seed trees have been retained but where little merchantable volume has been taken in the harvest. For many species, natural regeneration is not likely to occur or to become established after seeds germinate if brush or grasses hide the soil. Most kinds of trees that grow in even-aged stands require the mineral soil to be exposed in order for the seeds to germinate. These species usually also require full sunlight for survival beyond the seedling stage.

A third major problem involves the market. Will the owner be able to sell all the timber except the seed trees—everything from fencepost-sized stems to poles and pilings? If some trees are left behind uncut, the purpose of the system—to get a new even-aged forest—may be defeated.

Windthrow of seed trees also may defeat the purpose for which they were left. This is especially so if the species has shallow roots, weak tap roots, or if it typically grows in shallow, peaty or continually wet soils.

Erosion of soil and sedimentation of streams is more likely with seed-tree cutting than with practices that take less volume in the regenerating harvest. In wide openings, weed shrubs, brush, vines, briars and herbaceous plants often invade, preventing germination and growth of valuable trees.

Seed trees provided seeds for regenerating this southern pine stand. Note the magic circle surrounding seed trees. Seedlings do not survive within this zone that extends well beyond tree crowns.

What happens to the seed trees? If spared from lightning strikes, fire or other tragedy, they subsequently should be harvested and sent to the mill. If not, they compete with the new seedlings in a zone called the "magic circle." Within that circle, seed trees and seedlings for many species are mutually exclusive: both will not survive on the same site. The circle takes in much more than the area of ground covered by the crown. Seedlings will be affected for half again that radius.

If the new forest is to grow, the seed trees must go. Their sale should follow soon after the landowner is satisfied with the density (the number of seedlings per acre) of the new forest. Retaining seed trees a while longer may be appropriate as an insurance policy if there is some risk of wildfire running across the land. Seed trees will be among the biggest and best in the stand. Financial loss for the landowner occurs if they are not sold and hauled to the mill. That loss could exceed the cost of planting, had clearcutting rather than seed-tree regeneration been carried out.

The second cutting, removing the seed trees, raises the logging cost per thousand board feet (MBF) above that for clearcutting, but not near so much as for shelterwood and selection harvests. Management costs also run about midway between the clearcutting and selection harvest extremes. Skill and judgement is required in selecting the seed trees.

Species best suited for seed-tree regeneration harvests are the winged-seeded southern pines, ponderosa pine and Douglas-fir in the West, and red pine in the Lake States. These species are intolerant of shade, not climax species in ecological succession, and produce easily disseminated seeds. (A climax forest is a stand consisting of one or more species which, if the forest is left alone by man and no fires or tree-felling storms occur, will regenerate under its own canopy forever. A climax forest is self-perpetuating.)

Usually such shade-tolerant forests that regenerate well with seed-tree harvests are of one species, called pure stands. To regenerate mixed species stands by this method can be tricky. Trees of various kinds may not produce seeds simultaneously: one species may produce seeds every year, another every two years, and another every 10. Forests of mixed species occur because of ecologic and environmental relationships. Many factors, like fires, storms, past harvests and species' environmental and ecological characteristics play a role.

If adequate reproduction has already appeared in the forest under the older canopy and in openings between trees, landowners may consider another system more appropriate. The woods could then be clearcut, leaving the young seedlings to assure a successfully regenerated new forest. This is called advance reproduction. Or, if the understory has developed to intermediate size, a modification of the shelterwood method (discussed in the next chapter) or selection (see Chapter 8) might be used.■

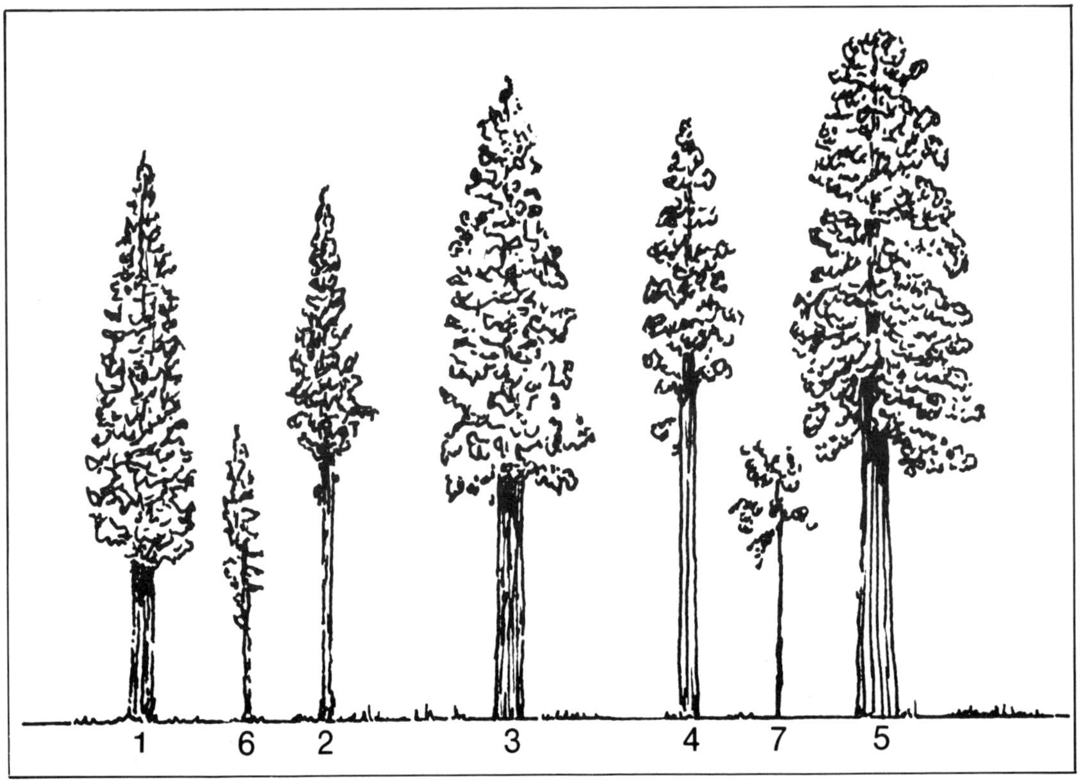

Dunning's classification for ponderosa pine helps to determine which trees to leave to provide seeds for the next generation. Trees in classes 3, 4, and 1, in that order, are the most prolific. Tree 1 is a dominant young or thrifty mature tree with good vigor; 2 is a codominant young or thrifty mature tree with good to moderate vigor; 3 is a dominant mature tree with moderate vigor; 4 is a codominant mature tree with moderate to poor vigor; 5 is a dominant overmature tree with poor vigor; 6 is an intermediate young or thrifty mature tree with moderate to poor vigor; and 7 is a suppressed mature or overmature tree with poor vigor.

CHAPTER 7

RENEWING THE FOREST—SHELTERWOOD HARVEST SYSTEM

Species of trees that grow in even-aged stands may be regenerated by the shelterwood system. As the name suggests, the method is appropriate for kinds of trees that require protection while stands of them are being reproduced. Eastern and western white pines provide an example of thin-barked trees which suffer sunscald when exposed to solar rays on the sides of boles facing west and south. Young saplings with smooth bark less than one-third of an inch thick are particularly susceptible.

The damage is not necessarily due to summer heat, but may be caused by either late afternoon summer temperatures hot enough to kill cambial cells (living tissues just beneath the bark) or by air temperatures dropping rapidly to below freezing after sunset in winter. Sides of trees facing north and east may be less susceptible because temperatures on those sides of the trunks do not fluctuate nearly so much. Injury takes place after growth of the summerwood (also called latewood, the last part of the growth ring) has been completed and before springwood (also called earlywood) has been produced the following spring.

This stand of a western pine lends itself to a shelterwood harvest, the first cutting removing about one-third of the stand. In logging, the soil is scarified, providing a seedbed for the seeds that are released from residual stems.

With the shelterwood system for regenerating forests on southern- and western-facing exposures, crowns of the residual stems provide protection for the new trees. The first cutting of the harvest removes about one-third of the stand (either by number of stems, volume or basal area—a measure of stand density). Gradually opening the forest in several cuttings permits early development of deep, corky bark layers so that, when complete exposure eventually occurs, the cambium ought to be immune to sunscald.

Diameters of openings less than the heights of residual stems permit sufficient shading from the south. Yet such openings must be wide enough for good seed distribution and seedling development. When the new stand is established, a second cutting takes out another third of the stand. Finally, when the understory has grown so tall as to be competing with the original forest for light, water and nutrients, loggers harvest the rest of the trees in the forest and send them to the mill.

Longleaf pines in the South often require a modified shelterwood system for stand regeneration. Here, the method is not so much to protect the seedlings from sunscald as to enable a seed source to develop and for the seeds to be dispersed while the overstory continues to grow. Frequently, longleaf pines come in naturally as advance reproduction under stands that appear as though they have received a heavy initial shelterwood cut.

To capitalize on this event, landowners utilize a variation of the shelterwood system. It works like this:

1. Thirty uniformly spaced, potentially productive seed trees on each acre are selected for subsequent stocking and insurance against later losses from fire.

2. These trees are released from the competition of trees immediately surrounding them in order to be stimulated to flower.

3. Just before seedfall, but a year before reproduction is desired, a third of the remaining stand is removed. The scarified soil from logging makes a good seedbed.

4. After an adequate stand is established, the rest of the overstory is harvested, taking care to avoid logging damage to the seedlings.

While southern pines are not usually sufficiently shade-tolerant to survive past the seedling stage on most sites in a shelterwood harvest, the method may serve to blunt criticism. Satisfactory reproduction establishment does occur on better sites; and the trees will persist where the second and third cuts can be made promptly as release of the young trees is needed.

The shelterwood system aids seed germination more effectively than the selection system by encouraging litter decomposition. This method exposes the mineral soil. Seeds for most species germinate more readily if in contact with mineral matter. Sunlight and warmth also stimulate germination and vigor for the seedlings springing from seeds shed from overhead trees.

Shelterwood harvests may take 30 or more years to complete, depending upon the species. For slowly growing, long-living species, like those in the Far West, landowners may wait 15 or more years between cuttings, while shortleaf pines in the South require but five years.

Stems chosen to be left after the first cutting likely will be sound, wind-firm trees of desirable species. They should be mainly in the intermediate and higher crown classes (mid-height to the tallest trees in the forest). Those retained for the final cutting should be sound, large and wind-firm, adding increment to valuable merchantable boles while awaiting their turn to be harvested.

It should be clear to users of the system that seeds and seedlings are anticipated only after the first cutting. Seedlings that germinate from a later seedfall may be a competitive hindrance to the originally germinated young seedlings.

While the system is especially useful for moderately shade-tolerant trees growing in even-aged stands, species that are shade-tolerant and which typically occur in a wide range of ages in the forest also may be regenerated by this method. Harvest times need not be rigidly established. One simply keeps an eye on the regeneration, releasing the seedlings as they need it.

Parts of large tracts may be treated simultaneously. Group- or strip-shelterwood harvests may also be appropriate. Openings made by groups are larger than those for single-tree choices. Shelterwood groups also take out more trees than do harvests by group selection (see the next chapter). Strips proceed with the cuttings in a direction against the prevailing wind, perhaps periodically moving into the forest 20 to 30 ft as the strip is enlarged and, at the same time, new seedlings become established.

Shelterwood harvests do little damage to sites. Soil erosion and weed shrub invasion are minimal. Cuttings, hardly noticeable to the passing public, have little negative esthetic effect on the forest scene. To avoid conflict over the esthetically disturbing appearance of tree harvests along roads, landowners sometimes use the shelterwood system.

Most of the disadvantages of the system involve economics. Much skill is required for the designation of trees to be removed in each cutting and for the felling and logging, lest the residual stems and new seedlings be damaged. Costs for management and logging are thereby high. Transporting the logging equipment to the same area three times also adds expense. But three cuttings will be much less than the 10 or so harvests required for the selection system (discussed in the next chapter).■

C H A P T E R 8

RENEWING THE FOREST—SELECTION HARVEST SYSTEM

Most esthetically appealing of the regeneration harvest methods is the selection system. Not to be confused with thinning, selection harvests have as their purpose the replacement of economically mature trees with young stems.

Typically with this method, some trees are removed every 10 years. The trees selected for harvest are soon replaced by seedlings in the openings made by the removals. Thus, with 10-year cutting cycles, each harvest removing 10% of the stand, the forest is entirely regenerated every 100 years. This period is called the rotation age of the stand.

Woodlands especially suited to selection-regeneration harvests are composed of shade-tolerant species which reproduce and survive well under dense canopies of parent trees. Those with this characteristic include the eastern and western hemlocks, the beech-birch-maple cover type of the Lake States and the Northeast, and the oaks and hickories of the eastern and southern forests. Many spruce and fir species of the West and North also reproduce well by this system.

Selection regeneration, as a silvicultural method, should not be confused with selective cutting. Selective cutting equates to high-grading, the logger taking the trees of highest economic value without any concern for continuing a healthy forest.

Landowners should distinguish between uneven-aged stands and many-aged stands. Selection harvests work well in, and produce, the latter. Uneven-aged stands may contain trees of only two ages. Such woodlands occur where the principal species are fairly intolerant of shade. In that event, seeds fall, they germinate, and seedlings become established. The new stock may appear vigorous, surviving for several years. Then, rather suddenly, the trees die.

Typical of this situation are several species of southern pines. Early in the life of the trees, tolerance to shade is rather high; later, seedlings of these species lose that tolerance. So it is that understory trees appear, grow and then die.

An open park-like stand in which much sunlight reaches the forest floor may be regenerated by the selection harvest system, periodically—over the rotation—taking out a percentage of the older and larger trees.

In choosing trees to leave and take, many characteristics must be considered. Species to be harvested must be marketable and tree sizes must be merchantable. Many trees in the stand may have no commercial value, possibly requiring individual treatment with a herbicide or by girdling in order to maintain the cutting cycle scheduled for the balance of the rotation. Consideration must be given as to which trees are seed producers, whether those to be cut are diseased, insect-infested, or ripe for an economic harvest. How they are spaced among their neighbors in the woods is another important consideration.

The managerial effort in choosing the trees to be harvested may be a costly investment for the owner. While buyers may be willing to mark the stand for harvest, this practice introduces a conflict of interest. It should be avoided; the landowner should employ a disinterested consultant to handle the task. Income from selection regeneration obviously will be less than for other regeneration-harvest systems.

The principal problem with the system involves logging damage to trees left in the stand. The cumbersome equipment required to skid, lift and move heavy logs inevitably scars trees and crushes seedlings. Logging scars serve as entry points for fungi, introducing rot. In this kind of harvest, especially, landowners may provide in their contracts with purchasers a bonus to loggers who do minimal damage to the residual timber.

The considerable skill required for logging adds to its cost; that, in turn, reduces the price buyers will pay for the stumpage. Low volumes per acre harvested and the need for workers to return periodically with heavy equipment to the same woods also add to the cost of the operation.

The biggest drawback, however, is the fact that, no matter how diligently a landowner and his forester may try, nature does not create a forest with tree sizes distributed according to the normal bell-shaped curve. There simply will not be the same number of small as large trees. The great majority of the stems in a typical woodland does not fall at the mid-point of the sizes represented in the forest. A perfectly all-aged stand with correct proportions of each age class (also representing a size class), from yearling-age to rotation-age, does not exist.

Landowners adjust for these vagaries by sometimes undercutting and at other times overcutting one or more of the periodic harvests. This forces the stand to be more normal in the next rotation and, therefore, more likely to provide a sustained yield. Decadent, suppressed, crooked, diseased, insect-infested, leaning and badly scarred stems must be removed, even if they are younger than the rotation age.

Group selection may be needed to regenerate some species of intermediate shade-tolerance in stands with size and age class distribution less than normal. With this modification, several trees close together are marked for cutting, regardless of kind, size and merchantability. Such openings enable seed germination and seedling establishment for species of trees not likely to survive and grow underneath the crowns of parent stems.

To determine tree ages in uneven-aged forests may require increment boring for annual-ring counting. The wounds made by this effort are entry points for fungi. Using increment hammers to determine current rates of growth also invites fungi infection. Spraying a fungicide into such injuries or plugging increment borer openings with fungicide-impregnated corks have been suggested as preventive measures.

The selection system lends itself to the "bruised-apple" syndrome. That is when the father every day sends a son to the fruit cellar to bring up an apple for each member of the family. Always the lad went with the parent's admonition, "Bring up the bruised ones first." Always when the boy picked over the fruit, freshly bruised ones appeared. They had shown no defect the previous day. The years went by, no one in the family ever enjoying a bruise-free apple. So it is with selection harvests: a landowner may be inclined to save the biggest and best trees, always catering to the superior stems and harvesting the sorrier ones, never knowing that all the while rot is creeping into the trees he is saving for a later harvest. Alas, the mill never saws a good tree from that owner's tract; and the owner never maximizes his potential income. To take some good trees is to "sweeten the crop."■

C H A P T E R 9

RENEWING THE FOREST—COPPICE HARVEST SYSTEM

Coppice regeneration involves the establishment of new forests by natural vegetative propagation. Dormant buds lying beneath the bark in the cambium layer at the bases of trees or in the roots burst to become new trees. Since the new trees do not arise from seeds, they cannot properly be called seedlings, although those that arise from roots some feet from a felled tree may appear so.

Most stands of broadleaf species and a few conifers may be reproduced by this system. The beech-birch-maple forest cover type of the Lake States and the Northeast lends itself to the method. So, too, the oak and hickory forests of the eastern and southern U.S., whether they grow on the ridgetops or in the river bottoms. In the West, aspen and the nitrogen-fixing red alder, which is becoming more important in forestry as old-growth trees are harvested, sprout when cut. Among the conifers that sprout are Virginia, pitch, and shortleaf pines. The redwoods of northwestern California are the only important commercial species of western North America that produce dormant buds which, when triggered by heat, break dormancy and grow into trees.

Cutting of an overstory allows light to get through the canopy to the ground. That light is converted to heat upon contact with the ground or with stump surfaces. Even a slight rise in temperature triggers the mechanism that results in buds developing into trees. Removal of upper tree parts may activate the buds by changes in hormone and carbohydrate concentrations that then take place within the growing regions of the trees.

These sprouts, also called root suckers and adventitious shoots, form near, or on, the tops of stumps. When that occurs, the resulting trees will be of little value. While the fungi that attack the dead wood of stumps, to cause their decay, are usually not those that attack living trees, the location of the new trees at or near stump tops encourages other kinds of fungus infection. Such trees also will be more subject to wind-throw, to breaking from the weight of snow, and may have a basal crook when reaching sapling size. Landowners may want them mechanically removed, since the effort may cause new and more sprouts to appear a little later. If herbicide treatments are used, they could cause death of valuable coppice. Tree-killing chemicals move throughout the living tissues of the tree.

While sprouts arising from stump bases (the root collars of the harvested trees) are superior to stump-

Coppice growth from a low stump in an oak forest. Neither the stems from stump tops nor stump bases make good trees. Merchantable stems arise from roots away from the stumps.

top shoots, those arising from along roots below the ground are preferred. The latter, however, grow more slowly than either stump-top or stump-base coppice.

All sprouts, regardless of place of origin, grow more vigorously than seedlings because they have available to them vast quantities of carbohydrates and nutrients stored in an extensive root system. Shoots arising along roots soon develop their own root systems which, in time, enable them to become quality trees. Aspens of the Lake States and the West demonstrate this tendency, as the new trees develop along the lines of the original roots as they radiate out like spokes from a dead tree or the stump of a harvested tree.

In addition to species, coppice regeneration planners must consider the size and age of trees at the time of cutting. Some kinds of trees that sprout prolifically in their early years fail to do so when old or large, the dormant buds in time becoming buried under a thick layer of bark which insulates them from the heat necessary to trigger their emergence.

Shown, too, is an oak sprout tree that is decaying, the fungi that cause the decomposition originating from the parent stump.

Thick, tough bark, as that of hickory trees, may also mechanically hinder penetration of a young stem, even if the bud has broken dormancy.

Because sprouts depend upon stored food in the roots, harvests of the overstory should be made during the dormant season. Reserves of starch and nutrients rapidly diminish in spring when the first flushes of growth, accompanying production of wood and foliage, consume the stored foods. Late spring is the poorest time to use a coppice system. At that time, the spurt of earlywood increment is complete and leaves are fully grown, but they do not yet actively replenish root reserves. For aspen, season of cutting seems to make little difference. This is so for both Rocky Mountain stands and for the "phoenix trees" of the Lake States that arose from the ashes of the burning holocausts and the cut-out-and-get-out harvests of that region's timber.

In using the coppice method, stumps should be cut low and on a slant in order to shed water. Stump decay is thereby minimized and sprouts are more likely to be low—near the ground-line root-collar, and therefore more desirable. Usually, timber cutters remove the entire overstory, for coppice regeneration of even shade-tolerant species requires full sunlight for survival and vigorous growth.

Short rotations are the rule for this method, even as short as 15 or 20 years. This means frequent harvests with consequent income. It also means the need to find markets for small, low-value products like fence posts, pallet lumber, stakes and pulpwood. Redwood is an exception to this general rule: the trees in the circles that surround stumps of the virgin forest also grow to old age and large size.

Unfortunately, only a portion—perhaps one-fourth—of the area of a redwood forest harvested by this method can be expected to be regenerated. Part of the blame is laid to the deer and elk that consume many of the new trees. So dense may be the sprouts from this conifer that thinning may be necessary.

Trees, especially conifers, that are able to coppice are important where fires destroy forests. Many stands of second-growth southern pine exist because flames killed the above-ground portions of shortleaf, pitch, and Virginia pines. In these forest types, too, thinning of coppice may be necessary if a forest useful for timber production is to follow.

All manner of wildlife take their toll of coppice

reproduction. Deer browse, mice girdle, and rabbits feed upon the youngest stems. Landowners may shy away from the method because of the inevitable esthetic insult. Appearance is much like that of a clearcut. To some it may look like rape of the land.

Minimal costs are involved in logging. Some skill is required after stand establishment to decide which sprouts to cull in the necessary pre-commercial thinnings.■

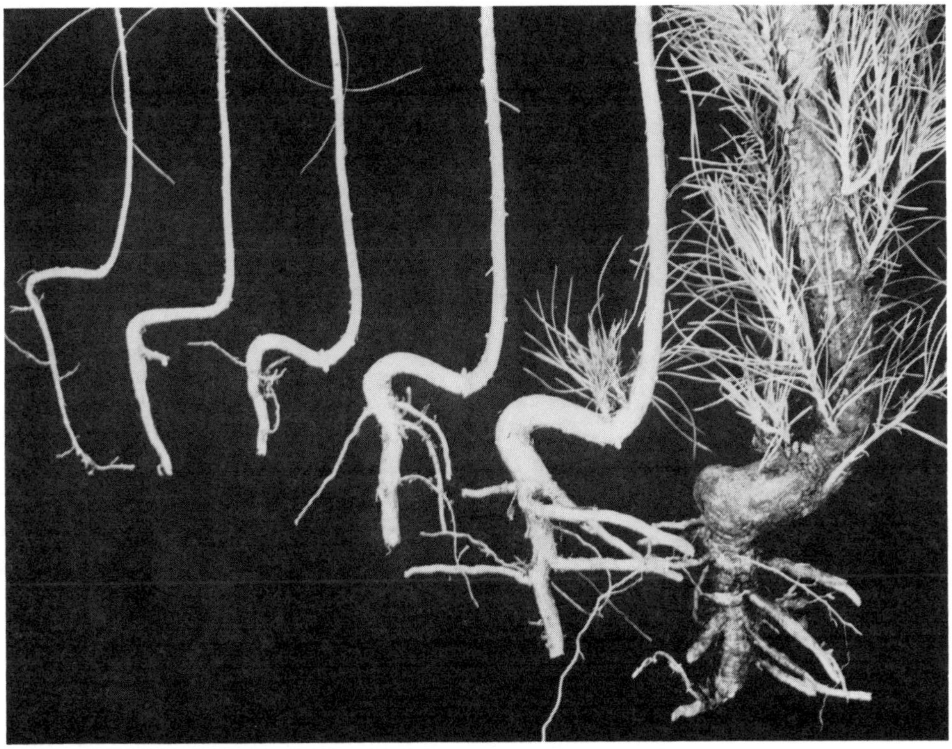

For shortleaf pines, coppice growth in burned-over forests develops from basal crooks, *Stone and Stone photo.*

C H A P T E R 1 0

PLANTING THE FOREST

As the forest becomes more noticeably important to the welfare of the nation, landowners will have a greater incentive to plant idle acres to forest trees. Even a financial incentive, however, is not enough for successful plantation establishment. For afforestation (planting trees on heretofore unforested lands) and reforestation (putting trees back on originally forested land) landowners need information about suitable sites for various species, planting methods and where to get the stock.

Many tree species are available from state forest service nurseries. In the South, these are slash, loblolly, shortleaf, and Virginia pines. Longleaf pine is rarely available. The first grows rapidly, has good form and makes good lumber. Slash pine often is susceptible to annosus root rot, a fungus that is difficult to control. Heaviest infections occur on lands formerly cultivated and abandoned from agriculture. The species should be planted in the southern extremity of the commercial forest zone. To plant it to the north invites ice storm damage. Preferred sites are moist flatwoods. These usually have rather sandy surface soils and subsoils of clay and clay loam that lie only several inches below the surface of the ground.

Tools of the planter: dibble or planting bar, planting mattock or hoedad, and seedling bag.

Loblolly pine, the South's most prominent tree, does well throughout the region and is recommended for the post oak belt west of the commercial pine-hardwood zone. It survives well on a variety of soils and soil moisture conditions. Hence, it can be recommended for both rather dry uplands and wet bottomlands in which hardwoods are ordinarily expected to dominate.

Shortleaf pine, more resistant than either loblolly or slash pines to ice damage, is the suggested species for the northern zone of the southern forest region. It is the dominant conifer of the Ouachita Mountains of Arkansas and eastern Oklahoma. It is suited to both hill and bottom sites, growing best in the more moist situations.

Virginia pine is typically an old-field "pioneer" (an ecologist's term for vegetation that first comes in naturally on denuded or cutover sites) species in the eastern states. It has adapted to a wide variety of sites, but should not be planted in low-lying, sometimes flooded, areas. Of the southern pines mentioned, Virginia is most likely to produce the best Christmas trees. Seeds of this species are sometimes available as genetically improved races developed for Christmas trees.

Just as genetics plays a role in grain and animal crops, so too is it important in forestry. Nurseries grow seedlings that produce fast-growing or well-formed trees and those with other characteristics such as high-density wood. Loblolly pines may be purchased that produce high- or low-density wood. Improved seedlings of this species may be selected for superiority in form of the bole, in speed of growth, or in drought-hardiness. The latter come from seed collected in the Bastrop area of central Texas, some 100 miles west of the commercial forests of that species. Trees grow there in rainfall averaging about 30 in. each year, in contrast to the precipitation that varies from 35 to 50 in. in the woods to the east. The Bastrop strain is especially recommended for high, dry hills throughout the region.

Shortleaf pine is available either from common stock, called regular, or from seed collected in areas of high-quality, mature trees set aside for cone production. Improved form and growth is anticipated, but not guaranteed, as the seedlings are first-generation offspring from the selected "wildlings" of the forest. Geneticists might say that one in four of the progeny should have the high-quality form of the parents.

Seedlings labeled superior in form and growth are grown from seed obtained in seed orchards. These costly tracts, in contrast to "seed production areas," are lands in which genetically improved stock has been planted and managed solely to facilitate gathering of high-quality seed. Often these parent trees are composed of branches taken from "super" trees (nature's best) and grafted to root stock of ordinary seedlings.

Orders for seedlings often are placed a year before needed. Many nurseries plant only the number of seeds of particular species to satisfy the order. Usually, at least 500 trees, minimum number for an acre, must be purchased for planting. If ordered from a government nursery, they may not be used for landscaping or for ornamental purposes. Trees are shipped collect, or picked up at nurseries, in bales or bags ordinarily of 1000 seedlings. Shipments in the South begin after the first frost and continue until March unless stock is exhausted earlier. In cold climates, planting is delayed until after the ground thaws.

Seedlings are packed for shipment in wet sphagnum moss to keep the roots moist and bundled in waterproof paper. Orders should indicate the planned planting time and arrangements made to do the job shortly after receiving the trees. Seedlings can be left wrapped a few days if kept cool and moist.

The accompanying chart shows the recommended method for planting, using the planting bar (also called a dibble), or a sharpshooter spade. The same idea holds for the hoedad planting tool.

Seedlings carried to the planting site must be kept moist continually. Even exposure of the roots to drying winds for a half-minute while removing the seedling from the bucket, inserting it in the hole and filling the hole can be damaging to the stock.

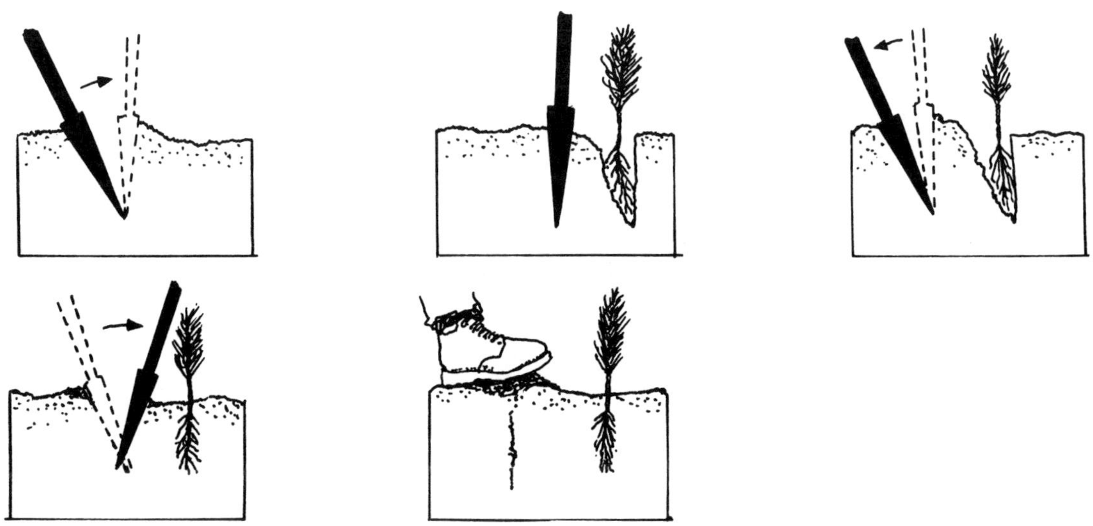

For hand planting, insert bar deep into the soil, then pull handle back toward planter to open the hole. Remove bar. Then remove one seedling at a time from seedling carrier. Hold by stem and needles to insert into hole to correct depth so that the root collar is an inch or two below the ground level. Insert bar 3 to 4 in. behind the seedling, now in the hole. Pull handle backward toward planter to firm soil around bottom of roots. Then push forward to firm the soil around the top of the root system. Repeat 3 to 4 in. behind the second insertion of the bar to assure against air pockets. Finally, pack soil firmly with a heel.

All of the root must be in the ground: any exposure will likely kill the tree. One cannot plant too deeply, as long as the terminal bud is above the ground. Placing the tree so that the root collar is an inch or two below the ground level assures against rain later washing away some soil, or wind erosion or soil settling that exposes roots.

Spacing varies, depending upon the product planned for the final crop. For general reforestation, 8x10-ft spacing, requiring about 550 trees/acre, is recommended. The 10-ft dimension allows pulpwood trucks and fire-fighting equipment to move between rows during thinning and fire-suppression efforts.

To accommodate pulp, waferboard and oriented strand board markets of the future, landowners may aim for 800 to 900 surviving seedlings/acre. For this stand, a commercial thinning in 15 to 20 years is anticipated.

For especially dry sites, and for those with deep soil, tube-grown stock is preferred. The trees are grown with a little soil in holders made of cellulose fiber, cardboard, or plastic. Organic containers are conveniently dropped into a hole made with a specially designed dibble or tool. Seedlings are removed from plastic tubes for insertion into the soil. With containerized seedlings, spacing can be greatly increased, permitting successful stand establishment with 250 seedlings/acre. Mortality is minimal with tube-grown stock, but the cost per seedling is high.

On many sites, preparation of the land with heavy equipment may be necessary. Landowners may wish to team up to hire this done.■

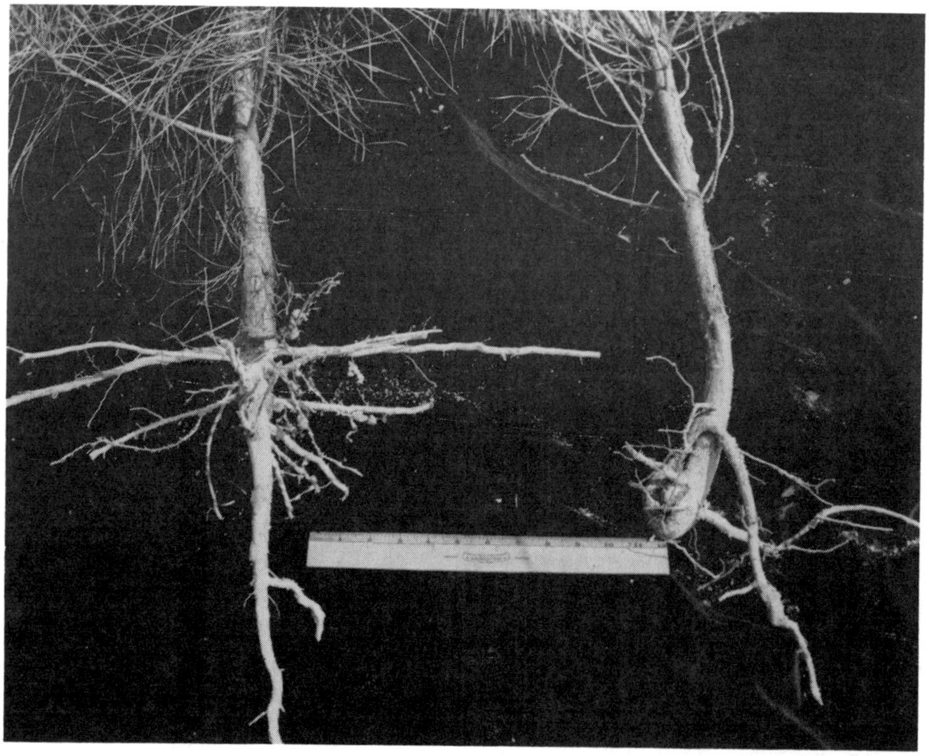

Planters need to be careful about root placement. The tree on the right will suffer, *U.S. Forest Service photo.*

CHAPTER 11

SEEDING THE FOREST

Considerable time and money can be saved by planting seeds, instead of seedlings, in the forest. Direct seeding, as the process is called, involves sowing seeds distributed by hand; by simple agricultural hand-held, crank-operated cyclone seeders; by powered ground machinery; or by aircraft. It is a means of artificial regeneration which produces a forest appearing much like that of natural regeneration.

Landowners may choose to collect seeds themselves from trees within their own woodland or from squirrel caches. This enables the continuation of the local and known provenance (locale of the source of seeds) for regenerating the forest. Purchased seeds may be from trees far distant, even 1,000 miles or more, from the area to be planted. As a rule, seeds should be collected within 300 ft in elevation of the site to be sown and within a north-south distance of 100 miles.

All species may be regenerated by direct seeding. Here walnuts are collected from an inferior tree for the purpose. This assumes that the poor form and lack of resistance to decay fungi are not inherited characteristics, *U.S. Forest Service photo.*

Some states have provenance maps to assist the seed collector, seller and buyer. California, Oregon, Washington, Michigan, Wisconsin, Minnesota and the Canadian Maritime Provinces publish them.

If landowners chose to collect seed, they must determine how to treat the species with which they are concerned. The pre-sowing process, called stratification, varies greatly. For some, true stratification places seeds in layers resting between strata of moist moss, the whole in barrels kept in cold storage at 35 to 41°F for one to three months. This treatment, typical of the procedure for pines, simulates nature's way: seeds fall in the autumn, lie in cold ground in the winter and germinate in the spring.

Other species are treated by soaking in dilute solutions of sulfuric acid. This technique (for junipers, for example) softens seed coats so that the seeds germinate shortly after falling to the ground, just as *Juniperus* cedar seeds do when passed to the ground from the acidic stomachs of birds. Other chemicals enable the seeds of a variety of species to break dormancy.

Another technique, also developed from observing birds, is the rotating of seeds in a drum (like a portable concrete mixer) lined on the interior walls with sandpaper. The rough-walled container scarifies seed coats as it whirls so planted seeds can readily absorb moisture. It acts like the grit-containing craw of a bird.

Seeds purchased from reliable suppliers will be tagged with provenance (place of origin), date of collection, stratification method, and type of repellent treatments. The latter may be especially important for some species in some zones. Longleaf pine seeds, for instance, are consumed in great numbers by mourning doves and rodents unless treated with chemicals to discourage the practice. Rodents and birds seem disinterested in small seeds, like those of western redcedar and Engelmann spruce.

Sellers of seeds should also be able to provide information on viability, sometimes called germinative capacity. This figure tells the percent of the seeds in the lot that can be expected to germinate within a certain period of time (perhaps 30 days). Buyers will want to know when the tests were made, for seeds lose viability with time. If requested, sellers should also be able to tell something of the vitality of a batch of seedlings. This subjective evaluation is based upon the appearance of the germinated seedlings in the viability tests. Should the laboratory seedlings in these trials appear weak, unable to compete and to extend their roots into the soil, the viability is considered poor. Landowners should increase the number of seeds per acre planted when that is the case.

Anyone can reliably estimate viability with a simple cutting test. Cutting through seeds with a sharp knife, the observer will note the percentage that are hollow or show rot. Again, if viability is low, more seeds must be planted to compensate.

Purity is also important. This figure tells how much foreign seed and trash (cone scales, wings, leaves, etc.) are in the lot. Seeds of a different size will foul distribution equipment. Pre-sowing treatment also involves getting rid of seed wings, husks and other outer coatings.

State and federal regulations apply to forest tree seeds, as they do to agricultural seeds, in many states. Labeling as disease-free may be a necessity if seeds are transported between states.

Tree-seed certification, the labeling of seeds as to the apparent genetic superiority of the parent trees, is available in many areas. Most seeds now commercially collected and sold by dealers are from designated "super" trees growing in seed production areas or produced in seed orchards.

A major problem for the landowner who collects his own seed is the harvest of seeds from undesirable phenotypes. (Apparent genetic characteristics describe a phenotype. All characteristics may not be genetically inherited; environment often affects the appearance.) Short-boled, open-grown trees often produce an abundance of seeds that are easy to collect, but the offspring from them will appear like the parent stock. Collecting seeds from high-quality, tall "super" trees in the forest is extremely dangerous. Only those properly equipped and trained should attempt the task.

Freshly collected seeds are stored in unsealed containers, lest the moist seeds be infected with killing molds even when temperatures are low. For long-term storage, seeds should be dried to 6% to 12% moisture and then stored in sealed containers at temperatures between 5° and 32°F.

The time for direct-seeding will also vary by species. While most tree seeds fall in the autumn and germinate the following spring, some require a whole year to break dormancy. On the other hand, longleaf pine seeds germinate within a month or so of autumn seedfall. As a result, direct-seeding that too closely imitates nature too far north in that species' range may find late fall seeds germinating by New Year's Day and the resulting seedlings killed by a January freeze. Yet a delay in planting the seed until after all chance of a freeze has passed could result in the seedlings not being sufficiently established to survive by the time the summer droughts—so prevalent in the species' range—arrive.

Distribution of seeds will also vary by species. The amount to be purchased (or collected) and distributed on an acre will depend upon their weight and how they had been treated. A pound of cleaned Douglas-fir seeds, for instance, contains 40,000 seeds; for redwood, 120,000 seeds; for longleaf pine, 5,000 seeds; and for eastern white pine, 25,000 seeds.

A rule of thumb is to disseminate about 20,000 sound seeds to an acre, anticipating 5% to 10% will germinate in a fairly well-distributed pattern. Pounds of seeds to be distributed on an acre can be determined once one has viability, vitality, purity and seed-weight information. The plunder of insects, larger animals (squirrels, raccoons, chipmunks, bears and most seed-eating birds), disease and weather will consume those that do not germinate. If germination is exceptionally favorable, providing too many seedlings, natural mortality or dominance by many individuals usually solves that problem. Some species, like lodgepole pine, express dominance poorly, if at all. Pre-commercial thinning then may be necessary.

More precise seeding may be attained by dropping the seeds in prepared spots, rows or strips than by broadcasting with hand- or cyclone-seeders. One man can seed about 20 acres of open terrain in a day with a cyclone seeder. The canvas hopper holds 5 to 10 lb of seeds, which the whirling blade slings out in a circle with a diameter of about 30 ft.

Other agricultural seed distributing equipment may be adapted for forest trees. From time to time, inventive people have developed mechanical seeders for certain tree species on certain kinds of sites. For large tracts, or ownership cooperatives of small groups, tractor-drawn rill and drill seeders used by farmers can be adjusted to accommodate most forest tree seed.

Small seeds may be dropped to the ground through 2-in.-diameter lightweight plastic pipes; the seeds are then pressed into the ground with the human planter's foot. Acorn dibbles that punch holes into the soil and then release the nut into the opening at an oblique angle are available.

Broadcasting by hand, from the ground with machinery or from either helicopter or fixed-wing aircraft does not allow for seeds to be firmly pressed into the soil. (Small forest owners may get together to contract for aerial seeding, even if their lands are not nearby.)

With hand-planting, survival is often improved if the seeds are placed on plowed berms, on beds of soil or on mounds made with shovels. This may be especially important if the site is wet. On the other hand, seeds placed in furrows will germinate and the seedlings survive better if the site is dry. At relatively little additional cost, seeds may be placed alternately on the berms and in the furrows, further assuring an adequate stand regardless of whether a dry or wet year follows. Tussock seeding involves the placing of seeds on clumps of organic matter, old logs and stumps. These micro-sites often provide good germinative ground.

Soil moisture at time of planting affects germination. If the site is moist, but well-drained, fewer seeds will be required. If the site is the extreme in either direction, more must be planted. Extremely dry soils could be planted with containerized seedlings rather than direct-seeded.

Grass sod makes a poor seed bed, although a one-year-old light cover may shield seeds from animal thieves and from the sun's desiccating heat. Seeds dry readily in sunlight, quickly losing viability and vitality. In this situation, workers use hoes or mattocks to scrape away the vegetation in order to bare the mineral soil. Fewer seeds will be needed on prepared sites, for there seeds readily make contact with mineral soil.

Landowners doing this work or overseeing it should be aware of a couple of safety matters. Chemicals used in treating seed to repel birds, insects and rodents may be toxic to human hands. Workers should wear gloves. If cyclone- or similar broadcast-seeders are used, workers should wear nose and mouth masks to avoid inhaling the dust of the chemicals as the seeds are propelled through the air. (Particular chemicals are not mentioned here because their labeling by the U.S. Environmental Protection Agency makes legal use of them subject to change.)

To determine success of a direct-seeding effort, the landowner tallies "milacre stocking." To do this, an acre is divided into 1,000 imaginary .001-acre square plots, each 6.6 ft on a side. The number of these on an acre which have at least one seedling is the milacre stocking. If 80 out of 100 milacres checked have at least one seedling (no matter how many more each milacre may have), the area is 80% stocked.

To readily find the milacre stocking, one walks straight across a direct-seeded field, aiming for a distant marker like a tree. With a long stride, each time one foot (left or right, but not both) touches the ground, it is in the center of the milacre. Looking 3 ft (3.3 ft, to be precise) in all directions from the point of the toe, the observer notes if a seedling is present. If so, the milacre is stocked; if not, it is unstocked. The milacre plots across the field lie contiguously in a line. They provide a sample of the tally for the whole area. The more such strip cruises, the greater the sample; and, also, the greater the reliability of the data.■

C H A P T E R 1 2

THINNING THE FOREST

Thinning a forest provides the landowner with the money to continue to manage his timber. Thinnings are also by-products of the forest that enable landowners to make a profit. Without this income, the bottom line on the ledger may bring neither joy to the owner nor coin to his purse.

Thinnings, often mistakenly called selection cuttings, must be distinguished from selection-regeneration harvests. The purpose of the latter is to get a new stand of trees of the same species established among the hold-over stems. Selection is one such technique. Thinning, in contrast, has another purpose: to enable the residual stems to grow more vigorously. Thinning removes competition, thereby providing more soil moisture, nutrients and light for the trees that are left to grow.

As a rule, when trees in a stand have live crowns less than one-third of the total height of the tree, thinning is needed. That much live crown is required to sustain and add wood increment to most tree species. After the stand is thinned, loss of lower limbs will not occur as rapidly as will height growth. The proportion of the tree with a live crown thus increases.

Stands, left unthinned, grow until crowding stagnates them. Adding some growth to too many stems keeps them in the low-value pulpwood class. Higher-priced sawlogs develop when growth is distributed among fewer trees.

Pre-commercial thinnings may be desirable. For some species, like longleaf pine, 10,000 to 20,000 seedlings/acre may compete for limited soil moisture and nutrients. Reducing the number to 5,000 with a sharp hoe, before brownspot needle blight takes its toll of the stunted stock, is recommended.

The landowner, preferably with advice from a forester, must decide how much of a stand to remove. This may be measured in volume (bd ft or

A pine stand needs thinning when little sunlight reaches the forest floor. A rule of thumb is to thin when the live crowns diminish to less than one-third of the total height of the trees.

cords/acre), number of stems per acre, or in basal area, another measure of stand density. (Basal area is the total area, expressed in sq ft, of the cross-sections at breast height of all the trees on an acre.) As a convenient measure of stand density, it combines the number of trees with the size of the trees on the particular area to tell the manager if the land has enough trees, too many, or too few to utilize the site fully.

A typical stand of southern pines may have at age 20 about 500 stems per acre. If these average 8 in. d.b.h. (diameter at breast height, 4½ ft above the ground), basal area will be a little more than 100 sq ft/acre. The volume of this forest may average 30 cords/acre.

Removing one-third of the trees in such a woodland is a frequent thinning recommendation. Thus about 170 trees, having a volume of about 10 cords and a basal area of perhaps 30 to 35 sq ft, would be marked for harvest.

Another stand, described as 50-year-old managed second-growth, may have about 160 stems, averaging 14 in., on an acre. Basal area would amount to around 120 sq ft/acre. Again, removing one-third of these trees would leave a few more than 100 trees and 80 sq ft/acre of basal area.

Of course, all the trees are not the same size. In fact, when marking for thinning, inferior trees, usually the smaller suppressed ones, are culled from the forest. Crowded oppressed stems, usually among the smaller ones, are also marked for harvest. Leaning and crooked trees are taken in order to enhance growth of the remaining individuals.

To speed up growth of the trees that are left may call for more drastic thinning than noted above. As much as one-half of the stand could be removed. On the other hand, if the forest is managed for highest-quality structural wood, light and frequent harvests are used. As a rule, the best southern pine fiber for strength and other mechanical properties is obtained when trees grow consistently six to eight rings/inch. (Similar rules of thumb apply for other species in other regions, but this is a useful near-universal generality.) To have this growth rate, the rule of thumb for southern pines is to thin for sawtimber on 10-year "cutting cycles."

Pulpwood growth, usually producing an increment of two to four rings for each radial inch of growth, is faster. For pulpwood, thinnings are generally made at five-year intervals. In contrast, growth occurred, or presently occurs, much more slowly in virgin tracts. In the primeval forest, trees of many species frequently grew, or grow, only an inch in radius every 20 or 50 years.

How often to thin and how much to remove may depend upon soil quality. Good soils—those containing optimum amounts of sand, silt, clay, organic matter, nutrients and moisture—grow trees rapidly. More frequent thinnings may be suggested for these productive lands than for sandy, sterile hills.

Trees to be harvested should be marked with paint that contrasts to the color of the bark of the species in the stand. Bright yellow is a typical choice. "Guns" are manufactured for this purpose. Some distributors market aerosol-propelled cans, similar to those used by vandals on public property. These are expensive. Oil cans like the ones that used to be used in auto chassis lubrication will do the job. So, too, will a paint brush.

Paint is applied at eye level, always on the same side of the tree so that the timber marker can see what has, and has not, been designated "to go." Another spot should be placed on the trunk close to the ground for convenient inspection following cutting to assure that only marked trees have been taken. The mark should be close to the ground to encourage wood cutters to leave low stumps. High stumps are not only wasteful; trucks get hung up, tires are damaged, and workers trip over them.

In young, dense stands, only trees to be left may be marked. This saves money. Some landowners like the "leave trees" left in rows, producing a plantation-like effect.

Cost of thinning depends on thickness of underbrush and accessibility of the site. Paint cost is minimal; some timber buyers provide it free for landowners whose trees are destined for their mills. Nor is the amount of wood to be removed important in determining cost. A round number—if the reader will take it with a grain of salt and not hold the writer accountable—is to allow one man-hour on the job for each two to three acres. Cost for marking may exceed the value of the stumpage sold, especially if fence posts are the marketed products. But remember, it is to improve the residual stand that the treatment is made. Indeed, there are times when pre-commercial thinning is desirable; the severed trees are left to rot for lack of a market.

Farm woodlot owners have been encouraged to use formulas for thinning their timberlands. A popular one is "D+6." The timber marker first measures the diameter of the trees in the stand and then figures the average. If, for example, the average is 8 in., the marker simply calls it "8," and to that 8, adds 6, giving the number 14. Fourteen ft, then, is the average distance the trees are to be spaced after the thinning. Trees in between these are marked for cutting.

Still another formula is "D × 1.75." Again, an average diameter of 8 in. is arbitrarily called 8 ft and multiplied by 1.75. This, too, gives 14 ft as the suggested distance between residual stems.

The chief advantage to formula thinning, a type of mechanical thinning, is that it can be used for most species on most sites by people with relatively little knowledge of timber values. This is also its disadvantage: the lack of that knowledge may result in quality trees, those which should be saved for the final crop, being taken. Some other stems of low value will be retained because they are not located at the proper points. Thus, formula marking removes the need for judgment.

And it is judgment that makes timber-marking fascinating silviculture for forester and landowner. It is an art.■

C H A P T E R 1 3

PRUNING THE FOREST

Not all the trees in the forest can have their value improved by pruning. It is important to the landowner to know which ones can be enhanced.

Early in the life of the timber stand, certain trees are chosen and permanently marked for the final crop. These crop trees—perhaps 50/acre—will have superior qualities: straight, fast-growing stems with small branches extending at right angles. Such branches leave smaller knots in the lumber than do branches of the same size extending up like arms lifted in prayer. Crop trees should also be free of disease and insect damage, suggesting the possibility of inherited resistance to fungus and insect attacks. If these superior trees are also able to produce viable seed, as noted by cones, fruit or seeds clinging to the trees' upper branches and lying on the ground beneath, so much the better.

Callus growth in a pruned upland central-U.S. hardwood tree, *U.S. Forest Service photo.*

In forests managed intensively, wood samples taken with increment borers from living trees may be tested for specific gravity. The higher the value, the greater is the density or hardness. Highest quality wood for structural purposes calls for dense texture and even growth—a steady six to eight annual rings/in. in radius. Finally, trees selected for this special effort should have healthy-appearing crowns.

Another criterion could be added to this list of "super-tree" characteristics. That is the ability to lose lower branches early. Some pine branches persist as long as 50 years: longleaf pine ordinarily retains them for 25 years. Aged redcedars and hemlocks may keep limbs all the way to the ground for the entire life of the tree.

For crop trees that do not naturally drop their branches, pruning provides clear, clean boles that command high prices for poles, piling, plywood and lumber. Ideally, the live crown extends down from the top of the tree no more than half of its total height. A crown one-third of a tree's total height is ordinarily sufficient for maintaining vigor and growth. No more than one-third of the live crown should be removed at a time, nor should the live crown be trimmed to less than 50% of the tree's total height.

Long-handled extension pruning saws enable workers to remove branches as high as 15 ft above the ground. Using an ordinary step-ladder may add another 5 to 10 ft of "clean" stem to the bole. As a general rule, pruning may be done to 8 ft at age 10 and to 16 ft 10 years later. This leaves a clean butt log.

Pruning may be done in two or three steps, cutting branches periodically before they reach an inch in diameter. The knotty core is thereby reduced, adding to the amount of clear wood in the butt log.

Large branches should first be undercut flush with the bole to prevent the bark on the trunk from tearing as the limbs fall. If, in the springtime, bark tends to tear, two cuts—the first 6 in. from the bole—reduces the risk. Then the downward flush cut is made.

Wound paints are not necessary for resinous trees, like most pines. In fact, these coatings may prevent the natural flow of gum that better seals the wound and thus protects it from fungus disease infection. Wounds usually heal in two to three years, depending upon their size and the vigor of the tree.

Pruning may be necessary for salvaging ice-damaged trees. Those bent more than 60 degrees from the vertical generally will not straighten satisfactorily without treatment. Drastic and prompt (within one month) treatment, removing most of the branches, lessens the weight on the stem. The tree will then tend to straighten itself.

Branches exhibiting fusiform rust on some slash and loblolly pines should be removed if the spindle-shaped cankers are within 15 in. of the bole. If this is not done, fungus strands will likely grow into the trunk, deforming it.

To do what nature has failed to accomplish because of a genetically inherited capacity to retain branches is why landowners prune. Even more than for structural softwood conifers, pruning is essential for obtaining quality hardwood logs, especially wild black cherry and black walnut, and to shape Christmas trees.

Broadleaf trees of high quality, and especially those free of knots, are used for furniture and panelling veneers. As plastics for vinyl flooring and synthetic fiber carpets become more expensive—because much petroleum is used as raw material in their manufacture—hardwoods may return to the flooring market. The wood of oaks, hickories, maples and birch could replace plastic furniture parts.

Pruning hardwoods is a skill. Special shears are used for small branches; regular pruning saws serve for those more than 1/2 in. in diameter.

For black walnut trees in the first few years of growth, branches are snipped several times during the growing season, beginning in the first May or June after planting. Under intensive culture, trees may grow so fast that excess branches ruin their shape. Pruning close to the trunk lessens the danger of forming loose knots. The year after a severe pruning, the trees need to be trimmed of small, soft sprouts that emerge at the wounds. (Pruned walnut trees, of course, have fewer nuts.)

Epicormic branching—sprouting along the bole of broadleaf species—may follow pruning because of increased light reaching the trunk of the tree. These small limbs may need to be removed, too, if tree value is to remain high. However, their persistence for a short period may give rise to a figure mark like that of birds'-eye maple.

For deciduous trees, branches are removed to encourage height growth and to keep crown width to a minimum, as well as to produce a clean bole. For them, a paint containing a fungicide is applied to the wound to prevent decay. Untreated wounds readily serve as sites for the germination of microscopic-size spores of rot-causing fungi.

Pruning may decrease diameter growth slightly. It may also reduce harvesting costs, lessening the amount of log trimming following felling.

Costs for pruning depend upon the number of branches to be removed and their accessibility. Most of the cost is for labor. One might generalize that a man-day should be allocated for pruning about 50 crop trees in an acre of forest.

Pruning trees in America's forests was once considered an important silvicultural treatment. This changed as the value of stumpage did not keep up with the nation's inflation rate and wood became more readily available due to wildfire control and the reversion of abandoned farm lands to forest. Pruning efforts virtually disappeared. As the cost of wood moves upward to accommodate the growing population and the greater demand for homes, fine furniture and panelling, pruning of quality trees for the final crop in the woodlands again will be done. Pruning the forest eliminates waste wood. It can be an important technique for maximizing the output of quality fiber for the future.■

CHAPTER 14

PRESCRIBED BURNING THE FOREST

Before 1945, intentional burning of the forests for any reason was taboo. That was about the time Smokey the Bear arrived on the scene to tell us so. His reminders are still with us. But about that time, too, foresters learned from observing nature that not all fires are bad.

Indeed, the pines of the South, the aspens of the North, and the lodgepole pines of the Rocky Mountains—three among many—cover the land today because of fires in an earlier day. Lightning and Indians effectively ignited the forest before the careless smoker, ornery neighbor, or faulty locomotive kindled the sparks that set the woods ablaze. While foresters still guard the woods from destructive holocausts caused by wildfire, they have many good uses for tamer fires that are carefully controlled.

Fires are prescribed for six principal reasons, all important in multiple-use management of the nation's forests. These are seedbed preparation, disease control, weed-tree control, hazard reduction, game-habitat improvement, and range improvement.

Table 1. Prescribed Fire and Alternative Suggested Costs.

Purpose of fire	Cost/acre dollars	Alternative	Cost/acre dollars
seedbed preparation	2	plow	20
weed-tree eradication	10	herbicide	25
disease control	2	fungicide	25
wildlife habitat improvement	5	food plots (cultivation, fertilizer)	30
range grass stimulation	5	fertilizer	100
hazard reduction	10	surveillance	?

Seedbed preparation fires, consuming grass and organic litter that cover the ground, allow tree seeds to reach mineral soil. Seeds of most pines and some hardwoods germinate only when in contact with inorganic earth. Burning for this purpose is helpful for both natural regeneration and direct-seeding.

Disease control is another reason for prescribing fire. Brown-spot needle blight of longleaf pine, in particular, requires such sanitation burns. Spores, found in the surface soil, are bounced by rain splash to needles. The disease, spread by those spores, infects grass-stage seedlings (naturally stunted trees resembling tufts of grass), killing the needles and slowing development. When low-intensity prescribed fire has consumed the spores along with the infected needles, the fire-hardy seedlings develop new, disease-free foliage. Two prescribed fires, at two-year intervals, may be necessary, the interim permitting fuel to build up to carry the second burn. Needle-blight, like seedbed preparation, calls for a fire of low intensity, usually a "cool" winter burn.

After a young stand of timber is beyond the sapling stage, foresters prescribe fires for weed-tree control. Undesirable stems—those of poor form or low-value species—are removed by hot, growing-season, fires. This allows the residual stems to be freed of some competition for sunlight, soil moisture and nutrients. Winter burns will kill back stems less than 2 in. in diameter. If roots also are killed, resprouting does not readily occur.

In the South, summer fires for three or four years in a row kill many rootstocks, reducing the sprouting of small hardwood stems by 50%. A single burn will likely be adequate, however, if pines are planted immediately—within a few weeks—following the fire. This early start gives the seedlings every possible advantage over hardwood sprouts which are bound to appear a little later. If this procedure is followed, about 75% of the planted pines are likely to be free-to-grow—not overtopped by brush sprouts—two years after the burn.

Flammable herbaceous material and accompanying brush should be eliminated, since it could be ignited by accident or arson. To avoid the blaze that might occur, hazard-reduction fires are prescribed. Depending on fuel and forest types, these may be cool fires, moving rapidly with the wind, or hot, summer burns driven against the wind. Prescriptions, of course, may call for intermediate conditions. Ground fuel in the southern forest consumed by these fires is called "rough."

Slash disposal is closely associated with hazard reduction. Here, though, the material to be consumed is the waste of logging—tree tops and branches—left in the woods. Prescribed burning also may be done to rid the forest of unsightly material, to clear the land for later mechanical planting, or to avoid a later wildfire under uncontrolled conditions.

Site preparation, a forestry term applying to preplanting efforts, is most inexpensively done by prescribed burning. This takes skill. The adage that "those who play with fire will ultimately be burned" is an admonition to be heeded. Prescriptions spell out the desired heat of the fire. Weed hardwood control requires hot fires, while site preparation of a grassy "rough" may be satisfied with a cool fire. Prescriptions also note the time of year and direction of spread. Fires running against the wind under certain prescribed conditions are hotter than those moving with the wind. Air temperature, time of day for ignition, relative humidity, fuel moisture and size of the area to be burned are other criteria a forester will likely note in preparing a prescription for a landowner. The owner should solicit a forester's aid.

[Site preparation removes vegetation which would compete later with the planted seedlings for soil moisture, nutrients and light. Full sunlight and adequate water are essential for early survival and growth of many species. Nutrients are generally adequate for survival: applying commercial fertilizer, however, often improves growth. Heavy equipment, hand tools and chemicals are other, and more expensive, methods for preparing sites for receiving seed or for planting. The method chosen depends upon acreage to be treated, financing arrangement, tools or labor available and the owner's preference.]

A controlled fire, in contrast to a prescribed fire, simply implies that it is not a wildfire and that stand-by crews are generally available. Controlled fires may give good results because of the inherent skill of, and years of observation by, the one handling the fire. Disaster may also occur. For either prescribed or controlled burning, most states require that the state forester's office be notified prior to ignition, lest the dispatcher needlessly send a suppression crew to the scene of the fire.

Game-habitat improvement and range improvement are associated reasons for prescribed burning the woods. An example of the former is to improve quail habitat in lower Gulf Coastal Plain forests. Careful burns remove the ground cover without in-

Prescribed fires are best set by personnel of state forestry agencies. Here, their people plow fire lanes and use a drip torch to ignite the fuel.

juring the trees, permitting new food plants to grow and the development of a lush, dense low cover in which birds can hide.

Prescribed fire improves the range for livestock. Burning oxidizes dry dead grass and broadleaf herbaceous forbs. In burning the organic matter, elemental carbon, hydrogen and oxygen return to the atmosphere as gases, while the mineral elements and nitrogen in the forage plants promptly leach into the soil. Fresh, succulent herbs then use the nutrients. Cattle know the difference. Burn a one-acre patch in the middle of a 100-acre pasture in which 10 cows graze: six weeks later, all 10 head will congregate on the small plot of nutritious forage.

How complicated is prescribed burning? Here is an example for a prescription for a longleaf pine seedbed preparation burn:

Wind: steady, out of the north, at 8 to 12 mph;

Temperature: below 60° F at ignition time, hence, a winter fire;

Relative humidity: 50% to 80%;

Precipitation: within 2 days of a rain of 1/2 in. or more;

Fuel moisture (water in leaves and debris on the ground): 6%;

Time of day: as early as ignition is possible, but not later than noon;

Direction of fire: with the wind;

Area size: 500-acre blocks; number of blocks limited by availability of standby suppression crews and equipment;

Manpower: one tractor-plow and three men for each block; and

Preparation: fire lines plowed, logistics determined, crews trained and informed.

Such a prescription should provide a fire so cool that no trees will be injured. In fact, if a grass-stage seedling happens to be on the land, its needles would be consumed without injuring the bud. From that bud, new vigorous needles will emerge. Someone has tested this by wrapping cigarette papers around buds. After a well-planned prescribed fire has passed, the papers were not even charred.

Fires running rapidly with the wind, under these conditions, are cool because they do not remain in one place long enough to build up much heat. Perhaps a dozen days a year may fit the prescription.

Prescribed fire is an important tool for multiple-use management. The technique is useful in all kinds of woodlands—private and public—to enable optimum production of wood, water, wildlife, range and recreation. While controlled fires are set to reduce snake and tick populations, this is not considered an appropriate forestry practice.

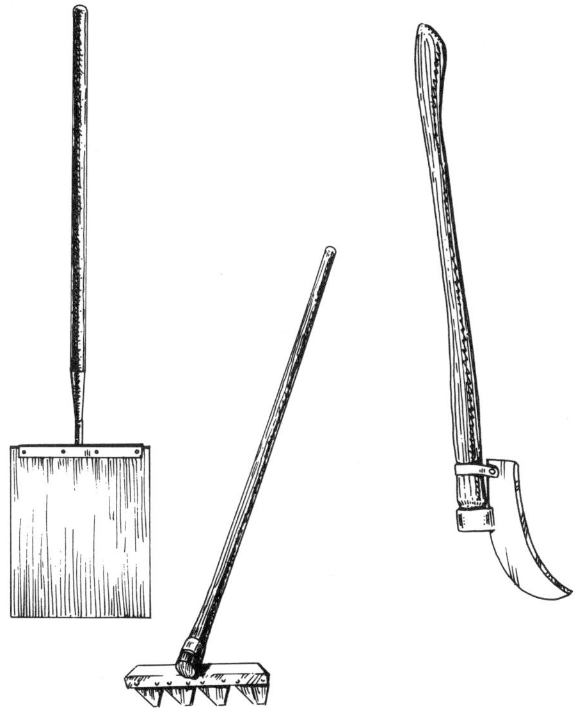

Fire flap for beating out flames, fire rake for making fire-lines, and brush hook for clearing lines.

Motorized equipment should be able to travel over fire lanes plowed along property boundaries and around interior stands separated by species. The lanes, when seeded to wildlife food plants, encourage certain game species to remain on the property. While making the lanes may provide a tax deduction, destruction of a stand by wildfire in some jurisdictions will not even give a break in *ad valorem* taxes.

Even the most proficient practitioner of prescribed burning may let a fire get away. Thus, the following paragraphs on suppressants and retardants:

Ammonium sulfate and ammonium phosphate are popular fire suppressants. Chemists for the Tennessee Valley Authority report that (1) ammonium phosphate crystal salts may melt or fuse to form a coating over the fuel, thus shutting off oxygen; (2) ammonia gas may be released through decomposition, again shutting off oxygen; or (3) a chemical reaction may occur between the retardant and the fuel, thus causing volatile fuels to become nonvolatile and, thereby, preventing combustion. Carbon dioxide, water and char result. When materials treated with ammonium phosphate solutions char, they do not burst into flame upon later exposure to high temperatures. Liquid ortho- or polyphosphate, diluted with water, is also useful.

Ammonium sulfate may be used in a dry form or in a water solution. Where clay or gum thickeners, iron oxide dyes and corrosion inhibitors are added, ammonium sulfate is marketed as Firetrol 100. Techniques for ground and aircraft application, published by the U.S. Forest Service, have been developed.

Borate (sodium calcium borate), a finely ground powder mixed with water into a slurry, provides long-term protection against fire. It is, however, no longer used in forestry because of its toxicity to plants (sterilizing the soil when applied in large doses), abrasiveness and mild corrosiveness.

What bothers some foresters is the chance that on some "just right" day, 50,000 acres may be ignited in forests with a prescription calling for a steady wind out of the north. Suppose on that day a warm front, carried by prevailing winds from the opposite direction, collides with the smoke-laden cooler winds. Such a squall line could bring rain to clear the air of the air-polluting particulates. But, because heat rises, the warmer air may instead move above the smoke-filled air, holding like a lid the cooler, smoky air near the ground. The probability of this occurring over cities, with consequent health problems, is unknown. This scenario is possible along the coasts of the South and Southeast.

Some states exempt prescribed fire from air control board regulations, with these exceptions: (1) such fires must be reported in advance to the appropriate agency; and (2) those doing the burning will be liable for injuries to people and property if visibility falls to less than two miles in rural areas and one mile in urban zones.■

CHAPTER 15

CONTROLLING SCRUB TREES MECHANICALLY

Weed trees have become a real problem in America's commercial forests. One big obstacle to profitable forest management is the low-value deciduous hardwoods which crowd out more desirable conifer construction timber species.

In most of the country outside of the West, but including some of that region, hardwoods are presumed to be the "climax" forest cover type. That is, if the woods were never harvested and holocaust fires and tree-tossing storms never occurred, broad-leaved trees would perpetually occupy the area. In the South and East, fire—natural or man-caused—often prevented the forests from attaining the climax even before the arrival of white men. Until the advent of the early settlers, however, fire apparently was not so prevalent in the Piedmont plateau; and, as recorded history indicates, most of this rolling country was forested with oak and hickory trees.

Beech, various maples and American chestnut were present to a lesser degree with the many species of oaks and hickories. Some shortleaf and loblolly pines were intermixed with the hardwoods, too. The soil in these deciduous forests was predominantly sandy loam underlain by red clay. Because of the fertility of that fine-textured subsoil, hardwood areas were preferred for cultivation and were rapidly cleared.

Row crops and erosion followed. So did the boll weevil and land abandonment. The result: two-thirds of the eastern United States reverted to "wild lands." While on many of these discarded farms, characterized by red clay at the surface, pines and hemlock have successfully seeded-in, so too have undesirable hardwoods. Even where pines got a head start, rapid advancement of now-useless hardwood species is apparent on areas stocked with conifers. This encroachment seems unrelated to the initial density of the pine stand: as many hardwoods come up in thick pine stands as in sparsely-stocked areas.

These weeds—trees, shrubs, and brush—have persistent tolerance to shade and a vigorous sprouting habit. Few markets are available for the fiber from these trees. Because of those three characteristics, over 50% of southern Piedmont land currently has a heavy hardwood understory. Another 25% will be added to the 50%; and land that would not grow cotton also will not grow worthwhile timber crops. The same discouraging word may be said of much of the rest of the forests in the eastern half of the United States. This situation prevails for extensive

Wide girdles are necessary to kill scrub hardwoods; narrow girdles callus over, the trees continuing to grow. Double-bit axes make the best girdling tool. One edge is more greatly tapered—for hard-bark trees like the hickories.

areas of young second-growth forest in the Far West.

Forest management systems may encourage this trend. For instance, where a suppressed hardwood understory exists, each thinning for pine paves the way for bigger hardwoods. After several harvests, well-advanced undesirable species may dominate the stand. To convert to pine is then a major feat. In pure conifer stands, early thinnings invite hardwood invasion because conifers are often essentially intolerant to shade: only on exceptional sites are abundant healthy pine seedlings found under full crowns of overstory trees.

The time lag between a harvest cutting and planting of adequate pine seed is another factor favoring hardwood encroachment. The ground surface shaded by deciduous trees usually doubles during the first two years following a harvest. Understory hardwoods encroach on 40% of the upland forest area, and one in every two acres reverts to low-value hardwoods following logging. To worsen the situation, Japanese honeysuckle and kudzu (both introduced from the Orient) and other vines cover millions of acres. In the western coastal states, add to those plants the chaparral and other scrub shrubs like manzanita, ceanothus and Scotch broom (the last-named brought by early settlers via the Northeast from Europe).

Aspect and slope influence invasion. Wetter slopes—the northern and eastern exposures—have greater numbers of hardwoods and, on these hillsides, the hardwoods become established to the crest of the ridge. In contrast, the drier southern- and western-facing slopes have considerably fewer hardwoods. Here, too, they are less successful in invading the upper reaches of the slopes.

To the landowner, this phenomenon is important if it may properly be attributed to the theory that southern and western aspects are too dry to support hardwood stands. In the absence of the broadleaved species, pines—naturally seeded and artificially planted—grow well. Apparently moisture is sufficient for conifer growth, but not for hardwoods. This is in contrast to the frequently accepted belief that broadleaf species are able to absorb moisture at the expense of evergreen trees.

There is no desire to eliminate all hardwoods. Only control is needed. Ample evidence has been presented that some broadleaf species are effective in improving the soil. About twice as much nutrient material is returned to the soil through leaf litter decay from hardwoods than from conifers. Hardwood leaf decomposition also improves physical condition of the soil. Increased porosity and aeration result when greater quantities of organic matter are "worked" into the soil; and prevention of the littleleaf malady, so prevalent in the Piedmont, depends upon that.

The oldest conventional method for killing unwanted trees is by simply girdling them. Removing a strip of bark and wood several inches wide *completely* around a tree will cut off all channels of food movement to the roots. It also will inhibit water movement from them. Callus growth curled inside hollows in the bark formed by rot and old scars must be completely severed.

In practice, trees under 6 in. d.b.h. are chopped down, and those larger are girdled by encircling with two rows of axe cuts and prying out the chips. Small trees can be removed readily with axes, machetes, bicycle saws, or hand-operated power saws. This should be done at the ground line to provide understory pine seedlings a head start over the inevitable sprouts. Sprouts originating from stubs waist high compete for sunlight with pine seedlings under five feet tall. Even where pines are less than two years old, cutting without chemical treatment will not be sufficient if the resulting sprouts overtop the seedlings.

Because young pines often outgrow sprouts from cut, or girdled, overtopping hardwoods, many areas need only one treatment. Within 10 years, the increased growth rate of stands so treated will more than offset the cost, and that is when the stand is beginning to produce merchantable wood.

Although over 60% of girdled hardwoods may be expected to sprout, few will shade understory conifers two years after girdling. Usually, less than 10% of the area shaded by parent trees also will be shaded by their sprouts.

There is less sprouting from the bases of smaller trees when girdling and felling is done in the late spring and early summer. During that period, carbohydrates stored in roots—the food supply for sprouts—are minimal.

The most efficient axe for this work is a 3½-lb. double-bit with a short handle (about 28 in. long).

The two-hack method is effective in reducing sprouting, although a longer time is required to totally kill the trees. Hacks are made 6 in. apart and the bark stripped from the zone between. By removing the inner bark, downward movement of carbohydrates is prevented. Because food reserves in the root are soon depleted, sprouting is inhibited.

Girdled trees die slower than those treated with herbicides. This may be advantageous if seedling release is to be gradual. Regardless of season of treatment, trees over 14 in. d.b.h. seldom sprout when girdled.

One way of determining the number of man-hours required to girdle trees with hand axes is to multiply the sum of the diameters (in inches) of all the trees by .005. Thus, trees totaling 200 in. in diameter can be girdled in about one man-hour.

A minimum of two man-hours/acre in stands of light undesirable hardwood density is required to do this job. Four man-hours/acre should take care of treating a stand of hardwoods of medium density that competes with pines for sunlight.

For large-scale land conversion from brush to sites prepared for pine seedlings, tractor-pulled undercutters, choppers, scalpers, shredders and root rakes are available. Such heavy equipment generally is used in sandhill and Coastal Plain areas where vast acreages of brush and grass are to be removed from level land. Unwanted trees, including their roots, are piled in windrows preparatory to winter planting of pines.

Another practice is to

1. reduce scrub oak cover in May or June, when root food reserves are minimal, using a brush cutter,
2. allow scrub oaks to sprout,
3. double plow during the summer to a depth of 10 in.,
4. allow oaks to sprout again to further diminish root reserves,
5. level area a third time, using a gang disk plow with 18-in. discs, and
6. then plant pines in the winter by machine.

Forestry consultants and contractors handle these tasks.

Minimum costs for this 6-point procedure may be as much as $100 (in 1986 dollars) per acre. The high cost of transporting equipment should be considered.

Where clay soils are cleared with heavy equipment and where discing or harrowing is not practiced, pine seedlings may not grow well. The weight of heavy machinery compacts the soil, which lessens aeration and water percolation to roots.

More primitive, but often quite effective, methods include skidding logs through brushy patches and dragging heavy cables or chains between two tractors moving parallel to one another.■

CHAPTER 16

CONTROLLING SCRUB TREES CHEMICALLY

The Second World War ushered in the use of chemicals in forest management. DDT had been discovered as the insecticide that would forever rid foxhole soldiers and prisoners of the fleas that carry the dreaded typhus germ. Inventive techniques for making fertilizer from cheap natural gas at the government's former ammunition plant in Alabama were accompanied by development of fire suppressants, fungicides and animal repellents. And chemists had synthesized herbicides like 2,4-D, intended to destroy enemy rice fields in the Orient and thereby encourage hungry soldiers and their political commanders to surrender. Herbicides were first used in forestry to an appreciable degree in the late 1940s.

Later, however, environmental lobbyists persuaded the United States Congress to control manufacture and use of essentially all chemicals introduced by man into the environment. By law, the label on a container of pesticide became a contract between buyer and manufacturer, or consumer and retailer. By that legislation, the user promises to apply the material in the recommended manner, as spelled out on the label, and to use the material only for the purposes approved by the Environmental Protection Agency (EPA). Hence, when a compound has been "labeled," EPA permits its application only as prescribed. The user is legally responsible for compliance. So, too, the seller has an obligation to know that the use of the chemical is within legal limits.

Forest landowners may encounter the word "registration." The Federal Pesticide Act of 1978 prohibits application of unregistered pesticides and also of registered ones in a manner inconsistent with label instructions. For some chemicals, applicators must be certified as being qualified. In several states, applications of chemicals on more than 10 acres of forestland may require inspection to assure the law has been upheld. Permits to apply certain chemicals may be issued by farm supply dealers. So complex are restrictions, and so frequently do they change, that the author and the publisher must disclaim all responsibility for claims by users.

Ammate, the trade-name for ammonium sulfamate and one of the first herbicides developed for forestry use, is effective for controlling undesirable broadleaf trees and shrubs. Scoop a tablespoon of the yellowish salt crystals from a wooden pail and pour it by hand into cuts, called cups, made at about 6-in. intervals at the bases of trees. Workers should wear gloves because they often discard the cumbersome wooden spoons for ladling the chemi-

Ammate crystals applied in "cups" cut into tree bases.

cal and use their hands. Ammate salt burns like table salt when in contact with skin that has been cut or scratched. And the hands of every woods-worker are scratched. Although wildlife often find the chemical in the cup an inviting salt-lick, no evidence has appeared to suggest injury to animals.

Ammate's corrosiveness prevents its use in metallic equipment. Water solutions are not necessary, however, because moisture in the air serves, within a few days, to dissolve the crystals. Labor-intensive application methods for ammate brought about the transition from this nonselective (kills all species) herbicide to other species-selective substances.

Among the first of the synthetic growth regulators for forestry use was 2,4-D, commonly called "D." For a while it was poured into frills (continuous cuts forming a trough) made with axes through the bark and around the trees at waist height. Then 2,4,5-T ("T") and combinations of D and T were employed in frill applications or applied as a basal spray to the butts of weed trees.

The first injector for herbicides, the invention of an Oklahoma lawyer named Little, was followed by other so-called "Paul Bunyan needles." The attorney-rancher knew nothing of forestry's need for the equipment; he was simply tired of fighting scrub oaks that invaded his rangelands.

Several modifications are now marketed. Injectors enable workers to place chemical solutions into

A tree-injector for applying a herbicide to tree bases.

A hypo-hatchet that injects a herbicide to deaden weed trees, *Texas Forest Service photo.*

the growing inner bark at tree bases. From there the chemical moves with the water flow to growing regions of the plant. These tools require much less effort than does the cutting of cups and the frilling of trees with axes.

Landowners also control brush or small hardwood trees in dense stands by using foliage sprays applied from mist blowers. The equipment may be mounted on trucks or trailers, or the chemical may be sprayed from helicopters or fixed-wing aircraft. When selective herbicides like D and T and 2,4,5-TP ("TP") (if the latter two should again be manufactured for sale in the U.S.) are applied as foliage sprays, pines are usually not harmed. Another popular material for foliar application is glyphosate (trade-named *Roundup*).

Herbicides are broadcast on the land in pellet form. The small solid spheres are dissolved by the moisture in the soil and absorbed into the roots. The chemicals move throughout the growing tissues of the tree. Some chemical pellets sterilize soil, possibly requiring years for the introduced compounds to break down, dilute and move from the area.

The hypo-hatchet, among the later tools to arrive on the scene, distributes concentrated chemicals to the inner bark through a hose connected to a small tank harnessed to a worker's belt.

Table 2. Some Herbicides Used in Forestry[1]

Common Name	Application Method	Selectivity	Relative Persistence in Environment	Hazard Rating[2]	Carrier	Trade Name (Partial List)
Cacodylic acid	Stem injection	Nonselective	Short	Caution	Water	Silvisar
Dalapon	Spray	Selective (kills grass, not pines)	Short	Caution	Water	Dowpon Dalapon
Glyphosate	Spray, injection, frill	Usually nonselective	Short	Warning	Water	Roundup
Dicamba	Spray, injection, basal bark, frill	Selective (kills broadleaf plants)	Medium	Caution	Water/oil	Banvel
Paraquat	Spray	Nonselective	Long	Highly toxic	Water	Paraquat
Diquat	Spray	Nonselective	Medium-long	Warning	Water	Diquat
2,4-D	Spray, injection, basal bark, frill	Selective (kills broadleaf plants)	Short	Caution	Water/oil	Weedone Esteron Weed-B-Gon
Ammonium sulfamate	Notch or cup	Nonselective	Short	Caution	Water (corrodes metal)	Ammate
Triclopyr	Spray, basal bark, hypo-hatchet	Selective (kills or injures all trees; grasses are tolerant)	Short-medium	Caution/ Danger	Water/oil/ kerosene	Garlon
Picloram	Broadcast, foliar spray, injection, spot-apply pellets	Selective (kills or injures all trees, most grasses are tolerant)	Long	Caution/ Warning	Water	Tordon
Hexazinone	Broadcast or spot apply pellets, spray liquid	Usually nonselective on brush and weeds. Selective on pines.	Medium	Slightly toxic	Various	"Velpar," "Velpar L" "Pronone" 10G
Amitrole	Spray	Nonselective	Short-medium	Caution	Water	Weedazol, Amizol
Atrazine	Spray	Selective (conifers not damaged)	Medium	Caution	Water	Aatrex, Atrazine
Simazine	Various	Selective (conifers not injured)	Long	Warning	Water	Princep
Fosamine ammonium	Spray	Selective (conifers not injured)	Short	Caution	Various	"Krenite"
Sulfometuron Methyl	Spray	Selective	Short	Very low toxicity	Water	"Oust"

[1] Courtesy Dr. Gerald Lowry *et al.*

[2] Hazard rating may relate to intake toxicity, skin irritation, eye irritation, etc.

Volatility of chemicals varies. The heavier the formulation, the less the danger of particle drift or vapor movement to nearby desirable vegetation. Some ester forms may be of relatively low volatility; amine salts are generally even less volatile. Invert emulsions (which look like black mayonnaise) find use in aerial sprays because little drift occurs. Constant stirring of non-invert mixtures is necessary because these compounds often form unstable emulsions when diluted with oil or water.

A frill made with an axe into which a chemical is poured.

Buyers should be aware that growth regulators are marketed in pounds of their acid-equivalent per gallon of chemical. Hence, a 1-gal container holding 4 lb of acid-equivalent provides twice as much herbicide as a 1-gal container with 2 lb of acid-equivalent. The word "acid" refers to the chemical description, not to the common conception of acid properties—such as burning and dissolving.

Forest landowners may ask about personal danger when using herbicides. T and TP are alleged to have caused human birth problems in Vietnam, where jungle areas were saturated with the herbicide to clear vegetation and thus protect American soldiers from ambush. Similar claims of human injury occurred in Oregon where weed-tree killing chemicals were used to release valuable conifers from competition. The Environmental Protection Agency's (EPA) dispute resolution conference of 1979 did not indict T. And the courts have indicated they would not have suspended the product on the basis of evidence available to EPA. Whether T or TP may harm future users is now a moot question: they are no longer manufactured in the U.S. Principal herbicides for forest use as of this writing are listed here.■

C H A P T E R 1 7

CONTROLLING INSECTS

Insects take more trees from America's forests than fire and loggers together. Industry and government have worked hard to solve the problems associated with the attacks. Trees have been felled, sprayed with lindane or benzene hexachloride (the once-popular BHC) and salvaged. This no longer occurs because BHC, of which lindane is a derivative, has been outlawed.

The worst in the South is the southern pine beetle (*Dendroctonus frontalis*). In the Rocky Mountains, it is the mountain pine beetle (*D. ponderosae*); and in the West, the western pine beetle (*D. brevicomis*). These insects hatch from eggs implanted beneath the bark. The larvae, about ¼-in. long, girdle trees as they feed upon the succulent inner bark. Later, the small grubs pupate, emerging as winged insects capable of flights of up to a quarter-mile to infest a new forest.

On occasion, several thousand acres may appear as though a fire crossed the land. Left in its "ashes" are the snags of large and small pines that served as food and shelter for the little black adult and its white larvae.

Evidence of the seriousness of the situation once led a southern governor to declare 34 counties disaster areas. The *Wall Street Journal* considered the insect predation national news, and the federal government allocated money for studies on how to control the pest.

Small forestland ownerships of commercial pine-hardwood forests are often harder hit than lands belonging to large industries. That is because the paper and lumber producers invest—on a per-acre basis—much more money for management than noncorporate owners can afford for that purpose.

The landowner's profit from forest management must come from "on-the-stump" sale of the wood he has grown. Industries—controlling about 15% of the commercial forests of the U. S.—earn profits from product sales. Standing timber is the raw material for those goods; without well-managed timberlands, the mills would shut down and stockholders would despair.

Perhaps in no way is the managerial care of the

Insects continue to destroy more trees than do fires in the U.S.

forest more closely related to catastrophe than it is from the outbreak and decline of pine beetles and other insects. Outbreaks most likely occur when trees are under stress, weakened by fire, drought, floods, storms and puddled soil. Puddled soil occurs when heavy logging equipment is used on clayey soils in wet weather, the compaction preventing adequate aeration of the ground.

Some woodsmen have noted that beetles are scarce following severe winters. It takes five days of 5 F weather to control the insect because the bark of pines is such good "thermal underwear." That kind of weather rarely occurs where the beetle and its hosts are found.

High temperatures also result in fewer insect generations each year and fewer beetles in each generation. Both extremes in weather delay developmental time, the period required for the beetles to grow from eggs to larvae to pupae to adults.

Tree color is good for diagnosis of beetle attack. "Flags" of red foliage appear in the tops of trees, allowing surveillance pilots to pinpoint outbreaks. But often the signals are too late. Many trees are already dead when the flags appear.

Chemical control is not the answer. The problem with chemical control is that insect predators are killed with the bark beetles. Some predators remain behind where infected logs are salvaged but are scarce in stands treated with insecticide. Because of this, some foresters recommend retaining a few dead, untreated trees in the woods to provide a home for the insect-eating predators.

Chemical control is now in limbo due to the Environmental Protection Agency's restrictions on the use of pesticides and the government's limited approval of the use of certain insecticides for specific purposes. For landowners to use pesticides which have restricted use for nonapproved purposes is to invite civil and criminal penalties of up to $1,000 and 30 days in prison under the federal pesticide acts of 1972 and 1975. (These are known as FIFRA, the Federal Insecticide, Fungicide and Rodenticide Acts.) The standards call for applicators to be certified as knowledgable of pest problems and control practices, of chemical use, storage and disposal, and of their legal responsibilities.

For the southern pine beetle, until insecticide use is approved, a cut-and-leave method at the smaller and inaccessible infestation spots is recommended. This is also the method to use where trees are below merchantable size or where there is no market for the wood.

With this technique, attacked trees are cut so that they lie toward the center of the spot. Then, noninfested trees at the "head" of the spot are cut, again so as to lie pointing toward the center. These harvested trees at the head—the direction in which the spot is moving—form a horseshoe-shaped buffer zone as wide as the trees are tall. Some foresters prefer to cut a buffer completely around the periphery of the spot. While some insects jump great distances to begin new spots, others move tree-to-tree until encountering an opening in the forest. These are good control methods. But an ounce of prevention is still worth a pound of cure. Thinning and otherwise reducing stand density is that ounce. Unfortunately, even this treatment is ineffective for the mountain and western pine beetles.

Dead trees with beetles no longer in them should be left standing. These provide food for *Dendroctonus* predators and parasites. Other insects, over 20 species known to be among their number, help to check beetle populations. Insect outbreaks encourage certain bird populations to increase, including the downy, red-cockaded, and pileated woodpeckers. Woodpeckers attracted by the *Dendroctonus* appetizer help control the beetles' numbers. With this kind of biologic control, the use of chemicals may be curtailed.

Another biologic control could be the blue-stain fungus. This pathogen infects logs between harvest and milling. The discolor confirming the presence of the pathogen is found in beetle-infested standing trees. Although the fungus is transported by the beetle, larvae do not fully develop where large areas of inner bark are occupied by the fungus. Consequently, blue-stain activity, increasing in abnormally warm summers, may partially but effectively control the southern pine beetle.

Biological control also calls for the use of predator viruses, often imported to an infested area for the purpose. Insect attractants, including sex pheromones, may find use in new control plans. To isolate potential attractants, the hind guts of thousands of insects were dissected under a microscope and a suspected pheromone carefully collected. When enough natural chemical was available, the material was analyzed and found to be an attractant. Chemists then determined how to synthetically produce it.

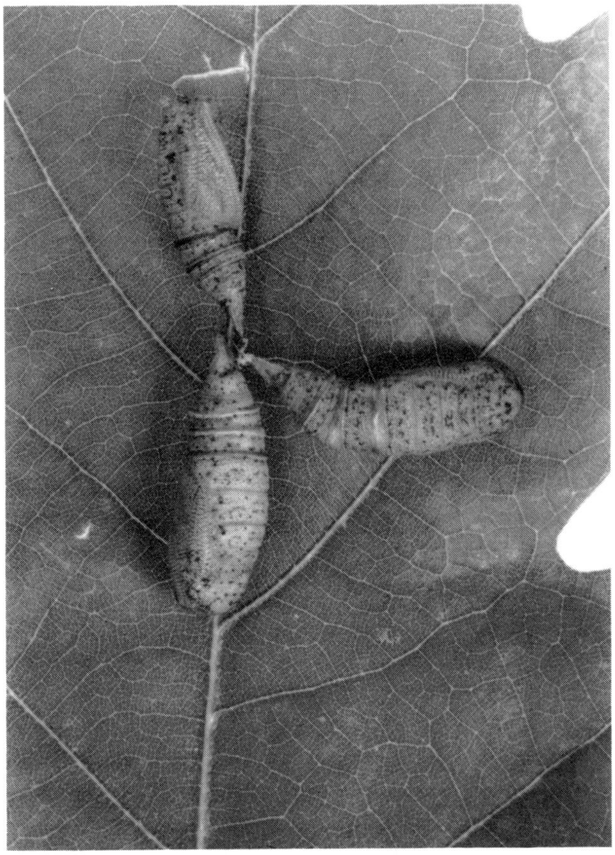

Larvae of the destructive elm spanworm, *U.S. Forest Service photo.*

One synthetic attractant, called "Frontalure" from the beetle's scientific name, is the basis for a new control approach. The compound is either spread on trees as a paste or placed in bottle caps which are then attached to trees. The attractant, like a bait, effectively holds the population to certain locales and restricts buildup of the beetle to selected brood trees.

Like boll weevils, pine bark beetles are "a-lookin' for a home." The beetle prefers large pine trees in dense stands that have relatively few hardwoods. Male and female mates may be the forerunners of eight generations looking for a home in such trees in a single year. Since each pair may produce 100 eggs, the offspring of a single pair could reach millions. From infested brood trees, most newly emerged adults move to adjacent trees in successive generations, laying eggs and killing trees as they radiate out from the initial host. Some, however, will migrate to stands of timber not previously attacked.

Attack it is! One report notes the highly disciplined army put in the field by this beetle. "It organizes and reorganizes its attack and supply forces seasonally, over half of the infested stems escaping detections by man."

If man cannot locate this forest forager, another insect can. It is *Thanasimus dubius*, a clerid (an antlike beetle) that follows and consumes the southern pine beetle. *T. dubius* does not mean doubtful in this case, for in the absence of chemical sprays, the insect is an effective soldier. Natural and normal control is a precarious balance between beetle and clerid, individuals dealing with individuals. Epidemics, those massive infestations that occur at ill-defined intervals, are associated with lag time in predator build-up following beetle population explosions.

Man finds the beetle evasive because attacked trees appear green, though already dead. Such stems cannot be detected by aerial observers and are difficult to find in walking through the woods. Infrared photography does bring out the infested trees because dead trees are a little warmer than living ones. The difference shows up on the heat-sensitive film.

First evidence of an attack visible from the ground is usually a dry, red, sawdust-like material, called frass, which collects in bark crevices at the tree base. Small entrance holes, the size of birdshot, can be seen on the bark at this stage. Pitch tubes, white or reddish globs of resin, emerge from some or all of the entrance holes, depending in part on the tree's vigor. Some larvae and parent beetles are engulfed and killed in the pitch flows.

Adults that attack trees carve tunnels in the inner bark. In these galleries they feed, girdling the tree as they do; and there they lay their eggs. Later, after the girdled tree is for all practical purposes dead, the eggs hatch to form small 1/8-in.-long larvae, or grubs. Although bark is an excellent insulator (trees would have frostbite and heatstrokes and be more readily injured by fire if it were not so), extremes in temperature do influence the vigor of the beetles inhabiting the zone of the living tree.

On the other hand, weather conditions which place trees under stress weaken them, making pine stems less resistant to beetle attacks. Floods and long droughts are examples.

To see what is going on within an attacked tree, landowners cut away the bark. For southern pine beetles, this reveals S-shaped tunnels, between bark and wood, from which smaller larvae burrows fan out, roughly at right angles. Foresters associate S-shaped tunnels with southern pine beetles, distinguishing them from the I-shaped tunnels of Ips, a less lethal species. (Associating the insects' initial letter with the shape of its tunnel helps to remember which culprit is at work in the woods.) The burrowing of the broods finally kills the tree by cutting off food transmission through the inner bark.

Egg galleries made by the western pine beetle exhibit characteristics similar to those for the southern *Dendroctonus* species. For mountain pine beetles, the egg galleries lie vertical to the long axis of the tree, terminating at their lower extremities with a slight crook. Boring dust fills the galleries except for their terminals.

Numerous holes without pitch tubes mark the emergence of the brood. By then the red flag of victory for the beetles is at the canopy tops. The needles are no longer green.

Thinning woodlands is now done rather infrequently. The cost to mark, sell and harvest a unit of wood and deliver it to the mill is too great to compete with clearcutting. This is because heavy machinery for small volume harvests is cost-prohibitive, tree-length logs cannot be skidded among the standing trees in the woods and each stem to be cut must first be marked by a knowledgeable person.

Many stems taken in the intermediate harvests, like thinning, are low in vigor. Others are cut because the stand is too crowded to sustain all the stems in it. Crowded and weakened forests—often the result of not thinning—are ripe for beetle attack.

Salvage of beetle-infested trees is appropriate. For the southern species, it must be carried out within a few weeks, before the fiber begins to deteriorate. Even before decay sets in, blue-stain fungus infects, thereby degrading the lumber sawn from salvaged stems.

Often asked is why the beetle was not a problem in the virgin forest? It probably was where stands were dense. But dense stands were not common in the woods of the South when English and Spanish pioneers arrived there.

Table 3.
Chemicals for Forest Insect Control

Elm spanworm	Dipel, Sevin	Spray on dry foilage when larvae are young
Pine tip moth	Guthion, Orthene, Sevin, Cygon, Pydrin	Sprays timed to coincide with early larval stages
	Di-Syston, Furdan	Apply granules to soil
Texas leaf-cutting ant	Methyl bromide	Soil fumigant (in winter only)
Pales and pitch-eating weeveil	Furadan	Root dip or soil application
	Dursban	Spray
Gypsy moth	Dipel, Orthene, Sevin, Methoxychlor	Foliar spray
	Disparlure	Controlled-release pheromone for mating disruption and population suppression
Pine bark beetles Southern pine beetle Engraver beetles (Ips) Turpentine beetle	Lindane, Dursban	Spray on infested logs (direct control) or standing uninfested trees (prevention)
Cankerworm	Dipel, Sevin, Orthene, Methoxychlor	Foliar spray
Forest tent caterpillar	Orthene, Dipel, Sevin, Malathion Diazinon, Methoxychlor	Foliar spray in early spring
Cone and seed insects	Guthion, Ambush, Pydrin	Foliar spray
	Furadan	Apply to soil

Trade names are listed here for user convenience.

Table courtesy of Herbert A. (Joe) Pase, III, Texas Forest Service

Today, 500 5-in. stems or 200 8-in. stems/acre frequently occur in plantations or in naturally seeded lands. In the virgin forest, 20 or 30 trees on an acre would have been typical. They may have been 200 years old, and likely 30 in. in diameter. Hence, the dense pine forests today provide abundant food for the beetle. Much food means many beetles, and that, in turn, means destructive outbreaks.

The woods are not silent. You can hear the steady whittling of the larvae as they convert woody tissues to their food, protected from the elements by the bark. In nature, it is the way of the woods. To control *D. frontalis* will require imitating nature. That means thinning stands before the beetles do.

As certain as the seasons recur, beetles and other insects will rise again to take their toll of the forests. The best protection from the gnawing pests is a well-managed forest. Care of the pineries *now* can better prepare these woods to sustain with minimal loss the next attack. But even well-managed stands may be devastated when lightning strikes weaken the resistance of host trees. Frequent inspections and early detection save stand losses.

Weevils of the genus *Pissodes* kill the terminal leaders in white pines and all spruce species. As a lateral branch takes over, the tree is severely deformed. This may happen to the same tree repeatedly until the tree is about 20 ft tall. At that time, the infested and deformed tree should be removed, allowing the residual stems room to grow and abundant moisture and nutrients to support them.

European pine shoot moths, brought from Europe before 1915, are especially damaging to red and Scotch pines. Stems are deformed as the larval grubs, hatching from eggs laid in needle sheaths and buds, bore into and carve out the insides of buds. Insecticides applied in the spring or summer are effective, as only one brood a year generally develops.

Pine sawflies infest many species of the northern woods, while budworms attack jack pine. Needle miners hollow out the interiors of the needles of jack, red, and Scotch pines, from the northeastern coast through the Lake States.

Gypsy moths, traveling with campers from one northeastern woodland to another since 1869 (when a French scientist introduced a new race of silk-producing insects), attack almost all broadleaf trees and garden plants to defoliate them. Yellow-poplar, locusts, poison ivy and honeysuckle vines, and sycamore are notable exceptions. Presently, control depends upon over 40 species of parasites and predators imported into the area. Some insecticides are useful. The principal symptom of attack is pinholes in the boles of the trees.

Douglas-fir provides habitat for *Dendroctonus pseudosugae*, a bark beetle. Inherited susceptibility to attack may depend upon the beetle: some slight difference occurs between those insects attacking the coastal variety of Douglas-fir and those feeding on trees of the interior.■

C H A P T E R 1 8

CONTROLLING DISEASES

Observations of the rapid growth of slash pine in the Southeast in the 1940s prompted testing it for purposes other than lumber. Paper manufacturing was growing in importance; only recently Georgian Charles Herty had discovered how to make newsprint from southern pines, including slash pine. The species became important as raw material for both brown-bag-style kraft paper and the inexpensive white sheet required for the daily paper.

Rapid growth of slash pine also led to its being planted considerably to the west of where it normally grows, far beyond its natural range. If this kind of wood made good lumber and paper in Georgia, it should do as well at the extremity of the southern pine forests in Texas.

But *Fomes annosus*, a fungus, was not anticipated. The fungus causes a root rot that destroys much timber in forest plantations. The problem is now serious wherever slash pines grow. The fungus is probably present in almost all soils. It is apparently most prevalent in agricultural soils or, at least, in those sites which have been plowed. (To call them "agricultural" may be to exaggerate their productivity.)

On such sites, abandoned by farmers because of land deterioration or for other economic reasons, slash pines were introduced. Planting was easy; most was by hand. Two-man crews with a dibble and a pail of 1-year-old stock could easily plant 2,000 seedlings a day on freshly abandoned farmland.

Blight damage to sweetgum, *U.S. Forest Service photo.*

The fungus of annosus root rot.

Survival was excellent, often exceeding 90% of the 1,000 trees/acre planted. Growth was usually better than that of any other species. Thinnings to remove four or more cords/acre were carried out as early as within 12 to 15 years to provide fiber for the new mills. Then something happened. After some stems were removed in these partial harvests, trees surrounding the stumps appeared weak. Foliage became sparse, needles turned slightly chlorotic, trees lost vigor and soon died. Infection seemed to radiate out from stumps, continuing to kill more trees with each passing year.

Spores of the fungus, so abundant in these abandoned fields, landed on the freshly cut stumps before the sticky resin could cover the surface. Those spores germinated, the fungus' string-like mycelia arising from the spore, growing down and into the roots. From the cut tree's roots, the fungus strands passed into the living roots of unharvested neighboring trees by way of grafts. Such grafts between trees of the same species are commonplace. (Never, to my knowledge, has a genuine root graft between trees of different species been discovered.)

The pathogen's attack is then underground from tree to tree. As the fungus destroys the roots, nutrient and water routes are lost. The tree dies.

Some novel techniques may control the malady. One suggestion is to fertilize stump tops with the urea form of nitrogen to stimulate growth of other, rotting-type, fungi that will hasten decay of the stump. More successful may be inoculation of stumps with wood-rotting fungus spores applied in liquid suspension to encourage stump decomposition before *F. annosus* growth proceeds very far toward the roots. Cultures for this biological control method, though still being researched, are now available. While *Peniophora gigantea* reduces losses, it will not eliminate *Fomes* infection.

How the attacking pathogen works is not clear. Probably the *P. gigantea* kills *Fomes* cells at the point of contact as well as decaying the wood from which both fungi receive their energy. Because *P. gigantea* is a sapwood-rotting fungus that occurs naturally in pine logging slash, some tree-disease specialists suggest felling a few trees before thinning harvests. This would build up the population of *P. gigantea* so that its reproductive spores will be abundant in the air during commercial cutting. Proper timing of the thinning is essential because fruiting of the fungus ceases when wood dries out.

P. gigantea cultures for annosus root rot control may now be purchased from a North Carolina firm. You can also get them, though not cultured, from the nearest stack of decaying wood.

Another biological control uses a *Trichoderma* fungus. This fungus neither eradicates nor competes with *Fomes*. It simply masks the latter by a rapid overgrowth. Hydraulic sprayers, mounted on the shears of mechanical tree harvesters to disseminate wood-rotting fungus spores, are being developed. Also being tested is a power-saw chain lubricant in which the cultures ride to the woods. As the saws cut stem from stump, some of the oil-pickled spores would drip to the freshly exposed wood and there proceed to decompose the cellulose fibers of the stump.

Instead of hastening decay, another control method aims to preserve the stump. To do this, creosote sprays are applied but, so far, with little success. Among other chemicals tested, borate holds much promise. The powdered material is dusted on stumps promptly upon the cutting of a tree.

Slash pine is not the only tree to be attacked by this *Fomes* fungus. All the southern pines are susceptible; and, farther north, white pine trees on abandoned farm lands and following partial cutting are often infected. Douglas-fir in California is seriously affected. In Scandinavian countries, the same fungus attacks trees, but, mysteriously, the result is a butt rot, rather than a root rot. The infection decays the inside of the basal four or five feet of a tree. When those cold-climate spores are tested in the U.S. South, root rot occurs. To add to the mystery, North American fungi introduced into Sweden produce butt rots.

Fomes conks, the fruiting bodies of the fungus, are brown on top and white beneath. They most frequently are camouflaged under needle litter at the base of infected trees or on stumps or roots. Sizes of the conks range from a thumbnail to a meat platter.

Timing is important for thinnings. Throughout the slash pine range, temperatures are too high in late spring and in summer for production of spores and for their growth following dissemination and inoculation. That, then, is the time of year recommended for thinning. Hopefully, further treatment need not be prescribed.

Other species are available which will grow almost as well, but without the risk of this disease, on sites selected for slash pines.

Owners of small forested tracts will encounter white pine blister rust, fusiform rust, and needle blights. The alternate hosts for blister rust, attacking white pines in the East, the Lake States and the West, are members of the *Ribes* genus of flowering plants. Gooseberries belong to this group. Only by destroying the *Ribes* plants can the pines be spared the infection. The task is laborious. The fungus *Cronartium ribicola* causes the malady.

The spindle-shaped canker describing the appearance of the infection of fusiform rust, caused by *Cronartium fusiforme*, and a similar, but spherical malformation, caused by *C. cerebrum*, evidences infection for most southern pines. Pruning branches with malformations within 18 in. of the bole usually prevents growth of the fungus' stringy mycelia into the trunk of the tree. If the spindle or sphere is farther from the bole than 18 in., infection of the trunk is unlikely. Infected trunks break at the site of the canker.

North America's most widely distributed pine, ponderosa pine, of which there are some six races, is host to many fungus pests. Rots in the boles and roots of living trees, attributed to several fungi genera, infect the wood through branch stubs, broken tops and fire scars. The boles of trees so weakened break or the trees die.

The resinous canker of white pine blister rust.

White pine root decline, caused by a fungus classified as *Verticiladiella procera*, is serious in the East in plantations in wet sites. Symptoms include delayed breaking of buds following winter dormancy and rapid death of trees. Fruiting bodies of the disease agent frequently occur in the galleries of insects on boles and branches of the infected trees.

Douglas-fir, the most important western commercial species, suffers from bark cankers (mostly from *Phomopsis*), needle casts (dropping off), rusts, galls and root rots. Perhaps more destructive is dwarf mistletoe, which is also severe on ponderosa and lodgepole pines. Mistletoe plants are seed-bearing, not spore-bearing like fungus-caused disease. Trees of any age in the forests of the Northwest may be deformed or killed as the semiparasite takes water and nutrients from the host tree while manufacturing starch for its own needs through the photosynthetic process.

Seeds are transported by animals, especially birds. They are naturally ejected with force many feet from a mistletoe plant in a tree to another tree or branch, where the seed germinates in a crack in the bark. The berry seeds, surrounded by a sticky gelatin, cling to the new host or to the bodies of the transporting animals. Deforming "witches brooms" of many branches occur at the place of infection. Infected branches should be pruned; infected trunks cut and removed.

Leaf casts, caused by *Hypoderma* spp. and *Lephodermium* sp. affect pines. Needle rusts also attack pines, disfiguring the trees, but with little effect on tree health.

Emergence tipburn, a needle blight of white pine of unknown origin, was first described in 1908. Growth is reduced and death sometimes follows. Symptoms include browning of current-year's needles from their tips, usually in midsummer. The dying terminal halves of the needles are at first reddish-brown; later, the dead portions weather and break off. Sometimes the whole crowns of small trees are affected, but this is not necessarily true of large trees.

In 1961, the disease symptoms became evident throughout the Southeast within several days. Foliar dwarfing, chlorosis, and stunted shoots may follow the needle dieback for several years, and it is accompanied by shriveled or wrinkled bark and raised pitch pockets. Extensive dying of feeder roots also occurs. The malady does not seem to have soil or tree-age preferences. Plantations and naturally occurring stands are affected. Varying degrees of tipburn among trees of susceptible species indicate that resistance may be attributed to genetic factors.

There is evidence that the malady is caused by atmospheric conditions, perhaps excessive ozone—the gas liberated by lightning and by the operation of electric motors, and a component of smog.

White pine is subject to a chronic tipburn, usually starting in winter and involving the whole needle. This condition results from atmospheric contamination by industrial wastes, which have not been fully identified.

Littleleaf, first reported in 1934, is a disease of shortleaf pine in the Piedmont province. The problem occurs only with the reforestation of severely eroded lands abandoned from agricultural cropping. Originally most farmland, although often underlain by heavy and poorly drained subsoils, had a well-drained A horizon deep enough to sustain tree root systems. With cultivation, however, the surface soil eroded to a thin mantle which forced roots to grow in the impervious, poorly aerated subsoils. Under such conditions, the littleleaf malady is found.

Although littleleaf does not develop solely from the presence of *Phytophthora cinnamomi* in the soil, the symptoms result from the gradual killing of fine absorbing roots by this parasitic fungus. Death of new root tips—those most susceptible to periodic killing—interferes with absorption of nutrients, especially nitrogen. Adverse soil conditions provide an environment which favors the pathogen.

Soil improvement, including fertilization, affords the best remedy for littleleaf disease. Before pines are planted on heavy clay soils, tilth and aeration should be improved by cultivation and incorporation of organic matter. Where existing trees show symptoms of littleleaf, fertilization and measures for improving aeration should be beneficial.

Rusts, caused by species of *Cronartium*, are a serious pest to many pines. The galls and cankers weaken stems so that they break in the wind. Diseased branches should be pruned if galls and cankers are within 15 in. of the main stem; beyond that, infection is not likely to spread to the bole.

Pitch canker on Virginia pine, caused by *Fusarium lateritium*, is recognized by a copious flow of gum far in excess of that accompanying any other known disease. The canker retains the bark and is usually sunken, the wood beneath being deeply pitch-soaked. Cankers which retain bark seldom produce spores. Beneath the bark are patches of light-colored or white fungus growth which turns pink or purplish and produces spores after the bark finally falls away. Twig cankers may form at the base of needle clusters or at injuries such as those caused by insect feeding. Trunk cankers develop on trees of any size, spreading until they girdle the trunk and kill the tree.■

Table 4. Fungicides Used in Forestry

Tree Species	Disease	Fungicide Treatment
Loblolly pine Slash pine	Fusiform rust	Bayleton Under high pressure, 1–2 ppa 3 times, from seed germination through June
Slash pine Longleaf pine (in seed orchards)	Cone rust	Ferbam, 1½ lbs. with 4–8 oz. spreader-sticker per 100 gals. water at 5-day intervals from time of fungus emergence until pollination ceases
Slash pine	Annosus root rot	Borax (fine granular) dusted on stump tops immediately after felling in thinning operations
Arizona cypress	Phomopsis infection	Benomyl (Benlate)
Southern pines	Black root rot	Methyl bromide (Soil fumigant) 300–400 ppa every 2–4 years in nurseries
Longleaf pine	Brown-spot needle blight	Bravo; bordeaux mixture (8-8-100); Maneb; Manzate "Benlate," seedling root dip

CHAPTER 19

DRAINING THE FORESTS

This is a story of water and trees—too much water for the best growth of trees. Its setting is the lower Coastal Plain of the Southeast. There, in the flatwoods, a bluish-gray waterlogged soil extends over vast areas. This clay loam is derived from thick beds of acid clay deposited at the bottom of the sea in earlier geologic times.

The soil is plastic and easily molded. Plodding through it on a damp day, one understands why they say of its tenacity, "You stick with it, and it will stick with you."

But when this soil dries out, it is extremely hard, shrinking with the loss of water, cracking with great crevices to make up for that lost volume and breaking into big angular chunks. Most soils like this occur near the coasts—just inward of tidal marshes—a few feet to as much as 50 ft above sea level from Virginia to Louisiana. Water frequently ponds to depths of several inches because of the flat or gently undulating topography with poor surface and subsurface drainage.

The impermeability of these fine-textured soils is an important factor in forest management. This characteristic results in only slight lowering of the water table at the edge of a forest adjacent to a clearcut area. In contrast, 35 ft within the stand, where transpiration of trees in the forest uses vast quantities of moisture, the water table is much lower. That is to say that water in the open fields, while nearer the ground surface than in the woodlands, does not move laterally to "seek its own level." Vegetation also affects the speed with which water tables are lowered during a dry summer period.

Much of this soil supports fair to good pine growth, yielding high-quality timber, pulpwood and naval stores. However, good pine stands are on the better-drained areas. On poorly-drained sites, pines are absent and low-grade hardwood species predominate. On these wet sites, natural regeneration of pine stands following clearcutting is difficult to obtain because water stands on the seedbed when seeds are ready to germinate. This problem is greatest in years with heavy spring rainfall. With clearcutting or seed-tree harvests, the demand by forest

"Bedded" loblolly pines planted on a drained site that formerly supported pond pine.

vegetation for water, transpired previously by the trees, is reduced so much that high water tables result. Death of residual seed trees may occur before a satisfactory new stand is obtained.

Woodland managers have long recognized the problem of wet sites. To many of them, it ranks as one of the most important obstacles to intensive management. While pines usually are restricted to better-drained sites, less-valuable hardwoods are dominant species in low-lying, poorly drained areas. Many areas could be covered with pine trees through natural seedfall if drainage were improved.

Drainage permits upgrading woodlands to better species because growth will be accelerated where excess water is removed. Fire control, tree harvesting, stand regeneration and other management practices may be performed more systematically if water does not stand on the land.

Ditch excavation, utilizing heavy machinery, constitutes a considerable investment. Whether the investment is justified by greater tree growth has not been fully explored. Nor has the amount of water been determined which should be removed for satisfactory seedling survival and optimum growth.

Deep-channel drainage lowers water tables two-tenths of a mile from the ditch. Height growth of trees is noticeably improved for half that distance. Frequently spaced shallow ditches would be more economical to install and equally beneficial to tree growth, especially where soils have relatively high permeability and where lateral movement of water is not limited. While few pine seedlings survive the first year when flooded to 8-in. depths for long periods, seedlings survive if the site is drained within a few weeks.

Temperatures of standing water on flooded plots average in mid-afternoon about 3°F more for 4-in. than for 8-in. depths of water. Summer temperatures range from 81°F to 105°F where the water stands 4-in. deep. Air temperatures at the same time have extremes of from 73°F to 97°F, water temperature changes always lagging behind.

Deep flooding of previously waterlogged soil, even for two weeks, may injure seedlings sufficiently to require replanting. Many trees make twice the height growth in the month from mid-October to mid-November as they do in the 4½ months prior to mid-October if the soil is moist, but drained. Complete dormancy may never occur in drained sites of the warm Southeast.

Here are some suggestions for establishing seedling stands of southern pines in southeastern coastal flatwoods.

1. Terminal buds must be above water to survive more than a few weeks. One-year-old seedlings slightly under 8 in. tall can survive a two-week inundation to a depth of their height. More than that and they die.

2. Flooding to depths of 4 in. may not cause heavy mortality the first year for southern pine seedlings, but heavy losses can be expected if the site is not drained by the end of the second year. Eight inches of ponded water causes great losses.

3. Excess water should be drained to depths of 4 in. below the soil surface for favorable growth of seedlings.

4. The rate at which flood water is removed does not affect survival. However, inundation as frequent as once in three weeks to a depth of 3 in. is too severe for optimum growth.■

CHAPTER 20

FOREST SOILS—THE HUMUS

Humus has an important influence on site productivity, watershed conditions and vegetation occurrence. It plays a role in ecologic succession.

Classification of forest humus layers is difficult. The problem relates to the loose, inconsistent and varying nomenclature and to lack of information on the origins and characteristics of humus layer types. It is "confusion worse confounded" because of the many timber types of varying ages, growing in innumerable soil types and regenerated under conditions ranging from eroded slopes to light selective harvests.

Generally, high summer temperature coupled with high humidity are conducive to rapid development of humus horizons, even when stands—natural or planted—are established on denuded sites. Temperatures sufficiently high for rapid oxidation of the organic matter in and on the surface of the soil occur in the warmer climates.

Hopefully this effort at description will not further increase the nomenclatural chaos described as "the seed of discord and confusion" planted by Ramann (who first recognized three humus types in 1893). In time, the system grew into an ideological and terminological monstrosity.

Humus-layer nomenclature probably originated in the latter half of the 19th century in northern Europe, where forests were and are more homogeneous than in the United States. There, pure stands of Scotch pine, Norway spruce or white birch cover vast tracts of sand plains of glacial outwash origin, moraine deposits, heaths or deep organic bogs. The rather abrupt transitions of these vegetation and soil types make humus-layer classification much simpler than in temperate North America.

Humus origin is principally influenced by vegetation, soil plants and animals, land form and geology. While a relation may exist involving time since disturbance of the vegetation, no such connection seems to occur between humus types and soil texture, slope or aspect.

Generally, humus types are either *mull* (consisting of mixed organic and mineral matter in the A_1 horizon) or *mor* (as when the organic matter is distinctly delimited from the mineral soil). The stratified layers of humus are L (undecomposed freshly fallen leaves, twigs, and other litter), F (a fermentation zone of decaying but still recognizable material), H (organic matter unrecognizable as to origin), and A_1 (organic matter mixed with mineral matter). (See Chapter 21 for mineral layer classification.)

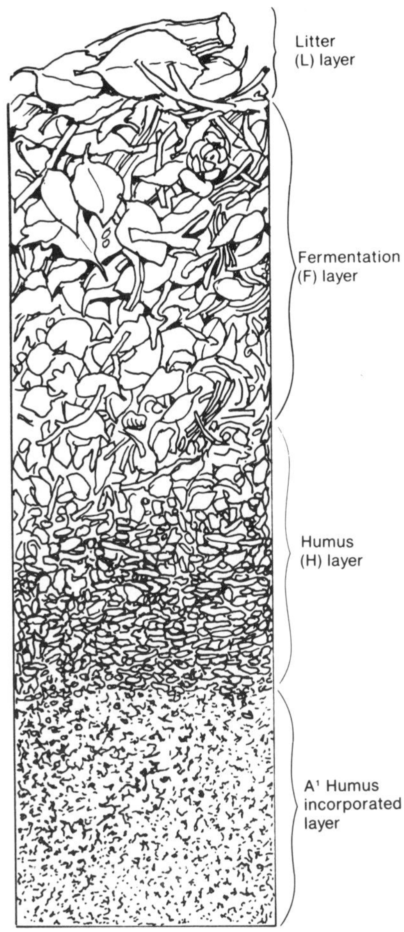

Principal horizons of humus in forest soils.

By definition, a *mull* humus layer consists of mixed organic and mineral matter. Less than 10% organic matter in the A_1 layer is insufficient for including a humus in a mull type, regardless of other morphological characteristics. Ten percent is a workable criterion in most climates of North America. This is an amount visible to the naked eye. True mor, in contrast, usually is without an A_1 layer, the transition from H to A_2 being rather abrupt.

Only mull humus types include A_1 layers and usually only mor types include H layers. However, due to the rapid oxidation of vegetable and animal remains, much organic matter is converted to soluble form and leached into the surface soil as well as remaining in amorphous form as an H zone. Humus is dynamic—constantly forming from plant and animal residues and continually being further decomposed by micro-organisms.

The firm mull classification is a dense, compact and structureless layer with less than 5% organic matter. In incipient humus, the L and F layers are present, the H absent, and the A_1 contains less than 10% organic matter.

The L layer in a mull humus type is generally not more than one year old. Usually such types occur under pure or mixed hardwood forests where mineral content of the fallen foliage supplies an abundance of nutrients for building up the population of earthworms, millipedes and lesser soil fauna which are active in foliage decomposition.

Pure white pine and hemlock forests of the East and West, in contrast, frequently exhibit a *mor* humus. The needles, being low in nutrient content, accumulate over several years and are finally decomposed by molds and other fungi, beginning at the lower edge of the L layer. An equilibrium may be established between litter production and organic matter decomposition for mull humus; but for mor types, the L layer, and perhaps the F and H, accumulate. However, under mull-forming conditions, the amount of organic matter in the A_1 is expected to increase, at least until climax vegetation is attained.

Duff is the "more or less" decomposed organic-matter layer just below the litter or, by another definition, unincorporated humus layer in a mor. *Duff*

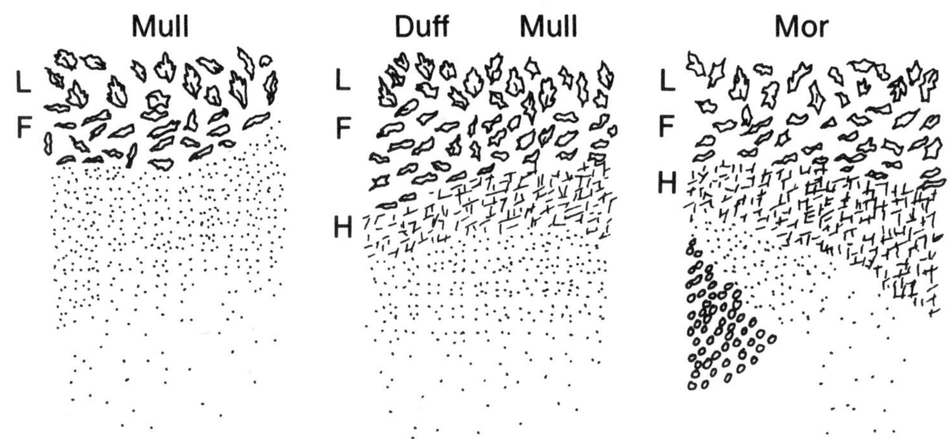

Kinds of humus layers in a forest: variations and combinations occur.

Forest Floor Layers

Litter (L) layer:	Unaltered dead remains of plants and animals. (Also called A_{00} and O_1).
Fermentation (F) layer:	Partly decomposed, but yet identifiable, remains of plants and animals. (Also called upper A_0 and O_2).
Humification (H) layer:	Well-decomposed, amorphous (but not in the sense of mud) organic matter. May contain some "worked-in" mineral matter (less than 10%). (Also called lower A_0; usually occurs only in mor types).
A_1 layer:	Melanized zone having an intimate mixture of organic matter (more than 10%) with mineral soil (usually occurs only in mull types).

mull refers to a humus type occurring in southern old-field pine stands. When the grass and herbaceous ground cover (called rough) die and quickly decompose, their remains become incorporated into incipient mull A_1 horizons.

Forest and humus types often are not highly correlated. For instance, wherever dense understories of ericaceous plants (like heather on the heaths) mask the direct effect which pines, spruces, and hemlocks exert on surface soil formation, the typical mor humus of conifer forests is not present. Leaves of hardwood trees, in contrast, tend to give rise to soil layers in which organic material and mineral matter are incorporated. Podzol profiles have an A_1 of incorporated mineral and organic matter, under which is an ash-white elluvial A_2 horizon. Iron, aluminum, humus, and colloidal clay are leached from this zone to form a darker B_2 illuvial zone where the chemical elemental sesquioxides of iron and alluminum, colloidal clay, and organic matter accumulate.■

C H A P T E R 2 1

FOREST SOILS—THE MINERAL LAYERS

Most soil is derived from disintegrating and decomposing rock and minerals. It must contain organic matter and be capable of supporting life. This medium for plant growth differs widely in chemical, physical, mineralogical, and biological characteristics because soil formation is controlled by climate, vegetation, parent material, slope and time.

Podzolization is essentially a process in which the *A* horizon is depleted of iron, aluminum, and other basic elements by rainwater leaching. These elements become partially deposited in the *B* horizon. Clays also move downward and are deposited in the *B* horizon. This process is active only in regions where rainfall is sufficient to produce trees. Soils so formed are quite acid as a result of the release of organic acids in the decomposition of organic matter. Podzolized soil generally has a sandy surface, composed mainly of quartz sand particles, while the subsoil contains considerable quantities of clay, also acid in reaction. Podzolization of forested soils often accounts for the depletion of nutrients essential to plant growth.

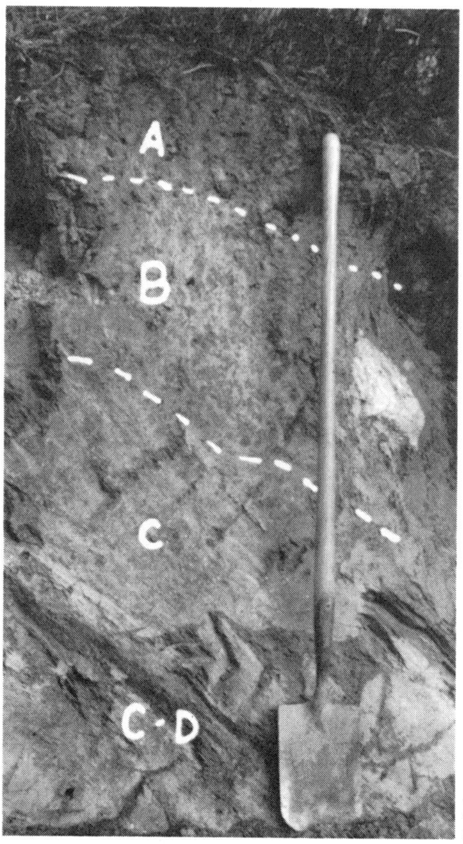

Mineral layers of a forest soil.

Soil *series* is an important unit in soil classification. It is defined as a group of soils having horizons with similar genetic characteristics in all respects except texture of the surface layer. It usually bears a geographical name (a county, river, etc.).

Type refers to soils similar both in genetic characteristics and texture of the surface soil. Size of particles, such as sand, silt, clay or their combinations, therefore, determine the type. The particular combination of these ingredients is referred to as *texture*.

Texture is often considered the most important criterion for forest soil classification. It can be used without reference to series. Site indexes are frequently based on this criterion. Landowners may obtain field kits for making these determinations or, better, become familiar with "feel" analyses.

In the field, texture is determined by the *feel* of the soil mass when rubbed between the fingers. The following suggests the obvious physical characteristics of the basic textural grades:

Sand: Individual grains can readily be seen or felt. Squeezed in the hand when dry, the grains fall apart when pressure is released. Squeezed when moist, the grains form a cast, which crumbles when touched.

Sandy Loam: Contains much sand but has enough silt and clay to make the soil sample somewhat coherent. Individual sand grains can be readily seen and felt. Squeezed when dry, a cast will be formed which will readily fall apart, but if squeezed when moist the cast will bear careful handling without breaking.

Loam: Contains a relatively even mixture of sand,

silt, and clay. Feels gritty, yet is fairly smooth and slightly plastic. Squeezed when dry, a cast is formed that will bear careful handling. The cast formed by squeezing moist soil can be handled quite freely without breaking.

Silt Loam: Over half of the particles are of the size called "silt," while a moderate amount of fine sand and only a small amount of clay are present. When dry, the soil is cloddy, but the lumps can be readily broken; when pulverized, the soil feels soft and floury. When wet, the soil readily runs together. Either dry or wet, casts may be formed that can be handled freely without breaking; but when moistened and squeezed between thumb and finger, the soil sample will not "ribbon," but gives a broken appearance.

Clay Loam: When dry, the soil breaks into clods or lumps that are hard. When moist and pinched between the thumb and finger, the soil will form a thin "ribbon" which breaks readily, barely sustaining its own weight. The moist soil is plastic and may form a cast that bears much handling.

Clay: When dry, hard lumps or clods form. These are quite plastic and usually sticky when wet. When moist soil is pinched between the thumb and fingers to squeeze out the water, the soil forms a long, flexible "ribbon."

Soil *phase* indicates a local feature. Examples of phase terms are: stony, cherty, hilly, eroded, etc. *Cecil sandy clay loam, eroded phase,* is an example of how soil series, type, and phase are designated.

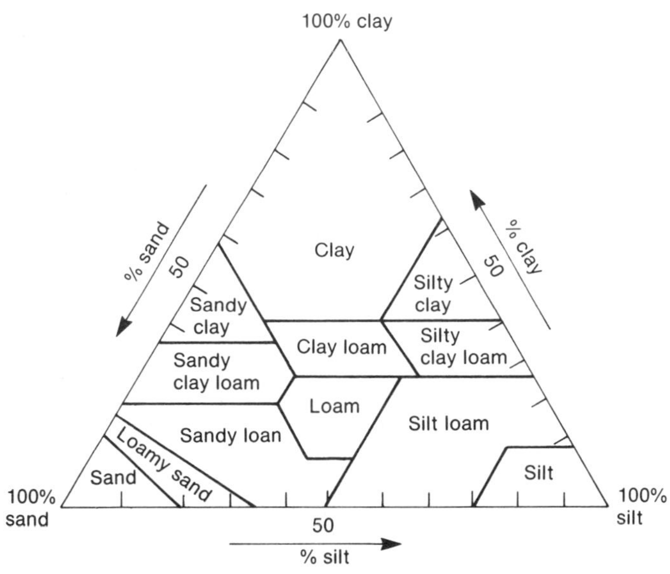

Triangle for determining soil texture.

Structure, the grouping of soils particles into aggregates, is important in forestry because this feature controls aeration and water movement. Structure occurs when mineral and organic particles are bound together to form plates, blocks, crumbs or granules. Why this "cementing" together of particles takes place is not clear; chemical reactions associated with organic matter decay and soluble mineral elements probably play a part. Earthworm casts formed from digestion and incorporation of organic and mineral matter are examples of crumb structure.

Sands in which no cementation has occurred are structureless. Such soils have poor water-holding capacity, but excellent rainfall infiltration characteristics. Well-defined structure generally implies good infiltration and water-holding capacity. Structureless clays, frequently the result of heavy machinery compacting wet soil, have poor infiltration properties.

Under pure pine stands, single-grained or granular-structured soils prevail. Low organic matter content in these mor humus type soils prohibits crumb-structure formation. The mull characteristics of hardwood stands encourage good structure formation.

Consistence is a general term combining the influences of texture and structure. Friable soils, at one extreme, may be sands or clay loams of crumb structure. At the other extreme occur plastic clay soils, which may be structureless or of platey structure. Definitions for consistency classes are:

Very friable: Coarse textured, loose when dry, coheres into granules when moist; sandy loam to loamy sand.

Friable: Breaks into granules when worked lightly between thumb and fingers while moist, but not wet; loam to clay.

Semi-plastic: Soil mass fractures when molded with fingers, ribbons to a certain extent when pressed between fingers and thumb; sandy clay loam to clay.

Plastic: Forms continuous ribbon several inches long when wet, aggregates are difficult to break with fingers when dry; sandy clay to clay.

Very plastic: Forms long thin ribbons when moist, cannot be broken with fingers when dry; clay.

In order to function as a medium for plant growth, soil must contain a certain amount of *wa-*

ter. Moisture as a solvent and carrier of nutrients regulates the physical, chemical, and biological activities of the soil. Soil moisture content also influences the length of time stomata remain open, and thereby influences the rate of photosynthesis.

Principal drainage classes are:

Excessively-drained soils: Very droughty soils with coarse, sandy or gravelly textures throughout. No mottling occurs. Soils with sandy or gravelly textures over gravel or bedrock are examples.

Well-drained soils: Adequate moisture except during the most severe droughts. Excess moisture from heavy rains is removed rapidly, yet enough is retained because of fineness of texture, depth to a moisture-restricting horizon, or other soil characteristics so that relatively high capacity for water-storage occurs within the root zone.

Moderately well-drained soils: Waterlogged at depths of 14 to 18 in. for short periods following heavy rains or for longer periods in the spring and fall. The waterlogging causes local reduction of iron compounds; the soil is mottled with gray and brown colors in the horizons where periodic high water tables occur.

Poorly-drained soils: Waterlogged to the surface for a larger part of the year. Only in midsummer or during periods of extreme drought is the water table more than 8 to 10 in. below the surface. For this reason, root growth of many plants is restricted. These soils are mottled throughout the profile—all the way to the surface of the ground. Apart from the mottling, the color is generally a dull gray. Colors below the surface tend to be dull, but in some sandy soils unusually bright colors and even a bright cemented subsoil may occur because of iron aggregation.

Very poorly-drained soils: Waterlogged to the surface a good deal of the time. Water stands for short intervals in wet seasons, but during the growing season the water table is generally just under the surface. These soils usually have peat or muck surfaces 2 to 10 in. thick and of black or dark brown color.

Essential chemical nutrient elements for plant growth are carbon, hydrogen, oxygen, nitrogen, phosphorus, potassium, calcium, magnesium, sulfur, iron, boron, zinc, copper, manganese, molybdenum and chlorine. The deficiency of any one may significantly reduce tree vigor and growth (see Chapter 22).

A discussion of soil chemical properties must include *pH*. A pH of 7 is neutral, a pH of 6, slightly acid. Adding lime and other bases raises the pH in the alkaline direction by reducing the relative proportion of hydrogen in the soil solution. Generally, hardwood sites are less acidic than pine, and forest soils more acidic than grassland soils. The maximum pH for a forest soil is a little above 8. That is a very sweet soil. At the other extreme, some swamp peat soils have a pH as low as 3. They even smell sour.

Decomposition of organic matter through the activity of soil flora and fauna, fire or increased soil temperatures—the latter due to opening up the forest to allow sunlight to reach the floor—raises the pH. Subsequent incorporation of released elements, particularly calcium, into the soil takes place as a result.

A sharp-shooter spade, soil auger, soil sampling tube (for surface cores) and a pH kit are useful tools for the forest landowner.■

CHAPTER 22

FERTILIZING THE FOREST

Abandonment of vast acreages from agriculture since the turn of the century and increasing demands upon the forests for lumber and pulp have inspired concern for intensifying forest management in North America. Since cellulose production is enhanced by nutrient fertilization of such fibrous crops as cotton and flax, it seems that forest trees, too, should benefit from supplements of inorganic chemical amendments.

Symptoms that indicate potassium deficiency in soils intensively farmed for about 100 years prior to the agricultural depression of the 1920s include:

1. needle chlorosis,
2. stunted seedlings and saplings,
3. short needles and
4. persistence of needles on branches for less than the usual number of years.

This situation is especially serious in deep, sandy outwash plains of glacial origin in the Northeast where white and red pines and spruce trees have been planted.

Soils derived under similar conditions of climate and geology also become magnesium-deficient for adequate growth of red, white, Scotch and jack pines. The most conspicuous symptom of this malnutrition is bright yellow discoloration of tips of current year's needles. Only under extreme magnesium deficiency is growth seriously affected. Treatment with $MgSO_4$ (Epsom salts), at rates of 20 to 50 lb/acre (ppa) of magnesium, alleviates symptoms and increases height growth.

Abnormal colorations indicative of potassium-deficient soils are observed in foliage of wild black cherry, gray birch and red maple trees and in the leaves of blackberry bushes. The presence of the abnormality in these species, considered weed plants growing in deep sandy soils, enables the landowner to determine whether the soil will be nutrient-deficient for valuable conifers without collecting and testing the foliage or soil. For black cherry, bright red pigmentation occurs at leaf margins, while gray birch has chlorotic margins sharply delineated from normal tissue. Red maple leaves are evenly chlorotic, and blackberry leaves display marginal browning.

All of these symptoms are removed by fertilizing with 200 ppa of potassium chloride (KCl), known commercially as muriate of potash. On sites displaying these abnormalities, potassium must be applied in, perhaps, the third or fourth year of the life of a

Gray birch leaves exhibit symptoms of potassium deficiency. Two hundred pounds per acre of KCl (muriate of potash) corrects the malady, as the leaf at the lower right illustrates.

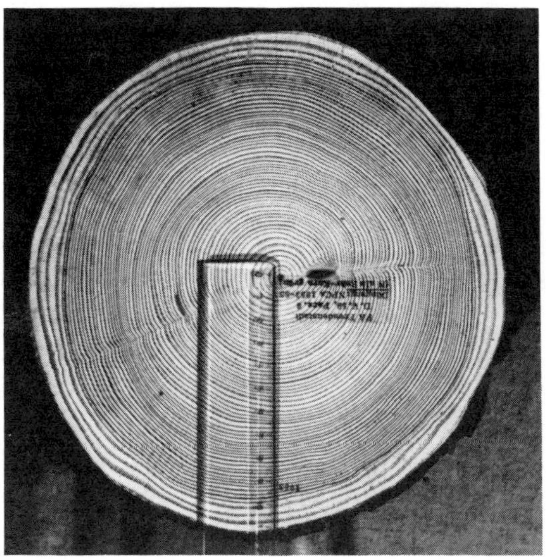

Note how nitrogen fertilizer stimulated the last four years' growth.

planted stand if suitable growth of demanding species is to be obtained.

Identification of nutrient shortages by leaf symptoms alone is not always a reliable diagnostic technique, as similar physiological conditions may accompany any number of nutrient element shortages. Deficiency symptoms for a single species on a single site may vary with seasons, and low levels of more than one nutrient may confound the symptoms. In addition, similar malformations often accompany insect and disease injury. As deficiency symptoms are only "symptoms," leaf and soil analyses should be used for more precise diagnosis.

Forest trees are known to exhibit nitrogen (N), potassium (K), phosphorus (P), and magnesium (Mg) deficiency symptoms under certain conditions. Perhaps the periodic chlorosis of southern pines is an indicator of one or more nutrient deficiencies. Frequently the chlorosis is consistent throughout the tree. At other times, it may occur on some needles only or may be apparent only on portions of needles. Nor is the chlorosis symptom consistent in its seasonal appearance. At times, iron and nitrogen have corrected, improved, or not affected this malady.

Nutrient-Deficiency Symptoms for North American Forest Trees

Nitrogen:	
Basswood	Yellow leaves; stunted growth; red petioles, stems and buds; first effect shows on older leaves.
Douglas-fir	Yellow needles.
Pine, loblolly	Yellow needles; short, stiff, yellowish-green needles.
Pine, longleaf	Needles pale green in lower half and light brown in terminal part.
Pine, shortleaf	Stunted growth; pale-green short needles.
Pine, pitch	Stunted growth; pale-green short needles.
Pine, Scotch	Pale yellow-green foliage with reddish-brown tips in primary and secondary needles.
Pine, white	Pale yellow-green foliage with reddish-brown tips in primary and secondary needles.
Locust, black	Yellow-green leaves with yellow to yellow-red veins and petioles; terminal buds die.
Phosphorus:	
Basswood	Dark green leaves; stunted growth followed by chlorosis and necrosis.
Locust, black	Golden yellow chlorosis on half of the leaf on the side of the rachis instead of normal reddish anthocyanin pigment in veins and petioles.
Pine, Scotch	Primary needles yellow-tipped and later reddish-purple.
Pine, white	Primary needles yellow-tipped and later reddish-purple.
Sweetgum	Veins and petioles red.
Pine, shortleaf	Short, necrotic needles beginning at base; stunted trees.
Pine, pitch	Short, necrotic needles beginning at base; stunted.trees.
Potassium:	
Basswood	Marginal and some interveinal necrosis of lower leaves.
Pine, loblolly	Needles turn brown, purple or yellow and die; top dieback; small necrotic spots on needles.
Pine, Virginia	Needles turn brown, purple or yellow and die; top dieback; small necrotic spots on needles.
Pine, Scotch	Yellow-tipped needles, eventually becoming light brown.

Pine, white	Yellow-tipped, abnormally short needles, eventually becoming light brown; needle persistence abnormally short; terminal shoots very short.
Pine, shortleaf	Blue-green needles, stunted growth; copper-colored terminal necrosis.
Pine, pitch	Blue-green needles, stunted growth; copper-colored terminal necrosis.
Cherry, wild black	Bright red leaf margins in mid-August well-defined from normal green pigmentation.
Birch, gray	Chlorosis of leaves, beginning at margins, in sharp contrast to normal green of leaf center (possibly also associated with magnesium deficiency).
Sweetgum	Margins of lower leaves bright red to light brown.
Locust, black	Margins of lower leaves bright red to light brown.
Pine, longleaf	Seedlings stunted; few, short needles bluish-green except at tips, which are dark brown.
Spruce, Norway	Needle chlorosis, followed by browning and finally death of needles; decreased height and diameter growth; decrease in the number of years needles persist; shortened needles.
Spruce, white	Needle chlorosis, followed by browning and finally death of needles; decreased height and diameter growth; decrease in the number of years needles persist; shortened needles.
Pine, red	Needle chlorosis, followed by browning and finally death of needles; decreased height and diameter growth; decrease in the number of years needles persist; shortened needles.
Maple, red	Leaf chlorosis.
Calcium:	
Basswood	Marginal necrosis extending to center of leaf, curling of leaf, later death of upper leaves.
Pine, longleaf	Pale coloration and delicate, weak needles; some needles brown at tips.
Pine, loblolly	Resin exudation from tip of bud before, or just after, bud breaking; death of buds; mottled yellow-green needle chlorosis beginning a short distance from tip, later progressing to tip; profusely spotted mature needles.
Pine, Scotch	Pale green needles in terminal regions with necrosis of terminal buds; some recurving of secondary needles.
Pine, white	Pale green needles in terminal regions with necrosis of terminal buds; some recurving of secondary needles.
Locust, black	Terminal leaves chlorotic, dwarfed, delicate; terminal buds die.
Sweetgum	Terminal leaves chlorotic, dwarfed, delicate; terminal buds die.
Magnesium:	
Basswood	Interveinal chlorosis.
Pine, loblolly	Yellow and brown needles (discoloration beginning from tip) and dieback; fascicular needles long and drooping.
Pine, longleaf	Tips of needles turn brown; seedlings chlorotic, appearing weak.
Pine, Virginia	Yellow and brown needles (discoloration beginning from tip), fascicular needles long and drooping.

Pine, Scotch	Primary needles blue-green with yellow tips; outer half of secondary needles long and yellow-green with golden to brown tips.
Pine, white	Primary needles blue-green with yellow tips; outer half of secondary needles long and yellow-green with golden to brown tips.
Locust, black	Light green scalping of edges of upper leaves.
Sweetgum	Upper leaves mottled light yellow-green between veins.
Pine, shortleaf	Tip chlorosis of needles.
Pine, pitch	Tip chlorosis of needles.
Pine, jack	Tip chlorosis of needles.
Pine, red	Tip chlorosis of needles.
Pine, white	Tip chlorosis of needles.
Sulphur:	
Basswood	Yellow leaves; early death of growing point, leading to lighter-weight stem and less leaves; shorter stem; shoot-root ratio reduced.
Iron:	
Pine, white	New needles yellow in terminal portions.
Pine, Scotch	New needles yellow in terminal portions.
Locust, black	Yellow mottling between veins.
Sweetgum	Yellow mottling between veins.
Pine, longleaf	Needles chlorotic, seedlings appear weak.

Nutrient Levels Below Which North American Forest Trees are Deficient (ppa = lb/acre; ppm = parts/million)

Nitrogen:	
Pine, loblolly	1.2% in foliage.
Pine, Virginia	1.2% in foliage.
Pine, ponderosa	3,000 ppa in soil.
Spruce, white	1.0% in foliage.
Douglas-fir	1.0% in foliage.
Phosphorus:	
Pine, loblolly	0.10% in foliage.
Pine, Virginia	0.10% in foliage.
Spruce, white	0.11% in foliage.
Potassium:	
Pine, Virginia	4 ppm in soil.
Pine, Virginia	0.28% in foliage.
Pine, loblolly	0.26% in foliage.
Pine, red	0.34% in foliage.
Pine, white	0.34% in foliage.
Spruce, white	0.15% in foliage.
Spruce, Norway	0.17% in foliage
Pine, white	0.50% in foliage.
Pine, white	20 ppm in soil.
Birch, gray	1.0% in foliage.
Maple, red	0.6% in foliage.
Magnesium:	
Pine, Virginia	24 ppm in soil.
Pine, Virginia	0.08% in foliage.
Pine, loblolly	0.07% in foliage.
Pine, red	0.16% in foliage.
Calcium:	
Pine, Virginia	3 ppm in soil.
Pine, loblolly	0.033% in foliage.

Growth of poplar trees is increased by applications of 300 to 500 ppa of a complete fertilizer such as 10-10-10. Materials applied deep in the soil give better results than broadcast applications. Water is more critical to tree growth than nutrients, even when the latter may be in short supply. The low nutritional demand of southern pines is also evidenced in the small amount of essential elements found in foliage—considerably less than in hardwood species.

Fertilization of hybrid poplar and yellow-poplar pays off on many sites. Diameter growth on the N-treated plots is improved in plantations 9 years old when broadcast-treated with 600 (ppa) of NH_4NO_3 (equalling 200 ppa N) and 500 ppa of 20 percent superphosphate (100 ppa P_2O_5).

Phosphorus, applied at a rate of 50 ppa of the available element, is generally effective for southern pines on sites deficient in this important element. Ironically, the deficiency sometimes occurs adjacent to phosphate-mining deposits in Florida.

Douglas-fir responds to nitrogen fertilization. A landowner can treat 50- to 90-year-old stands with urea at a rate of 100 ppa of nitrogen. Fertilizer pellets containing urea-formaldehyde resin and superphosphate, placed in planting holes, improve growth of seedlings in each of the first five growing seasons after planting. Survival of the seedlings may not be affected.

Most injury from fertilizer at time of planting has been due to excessive competition for soil moisture by weeds which readily respond to fertilization, rather than to salt injury.

Generally, where particular nutrient deficiencies are unknown, high-analysis complete fertilizers, such as 12-12-12, serve best.

Unless potassium is supplied to the areas of glacial outwash, referred to above, "vegetative retrogression" occurs. That is, planted spruce, white pine, or red pine give way to poverty grass, which in turn is replaced by *Polytrichum* moss, and then the less-demanding mosses and lichens. Finally, all vegetation disappears and erosion occurs.

Fertilizing with 200 ppa of muriate of potash reverses this process, ecologic succession replacing retrogression. Opportunities for effective fertilization elsewhere are usually less dramatic and the limiting nutrient element or elements more difficult to ascertain.

Flowering and cone production of slash, loblolly, and longleaf pines are increased by fertilization. Nitrogen, potassium and phosphorus are involved, but it cannot be said which element does the most good. Soil or tissue tests are required.

Fertilization may influence insect attacks, especially for Nantucket pine tip moth in southern pines. Fusiform rust infection seems to increase: supplemental nutrients encourage early breaking of winter dormancy. This, in turn, provides succulent needle tissue through which fungus spores enter at the very time spore dissemination is most prolific. On the other hand, nitrogen fertilization improves shortleaf pine trees affected by the littleleaf malady. This disease is associated with the fungus *Phytophthora cinnamomi* on severely eroded sites which once were in row-crop agriculture.

Conventional fertilizers, solid or liquid, are generally applied in forests by hand or by air broadcasting. There appears to be no advantage to drilling

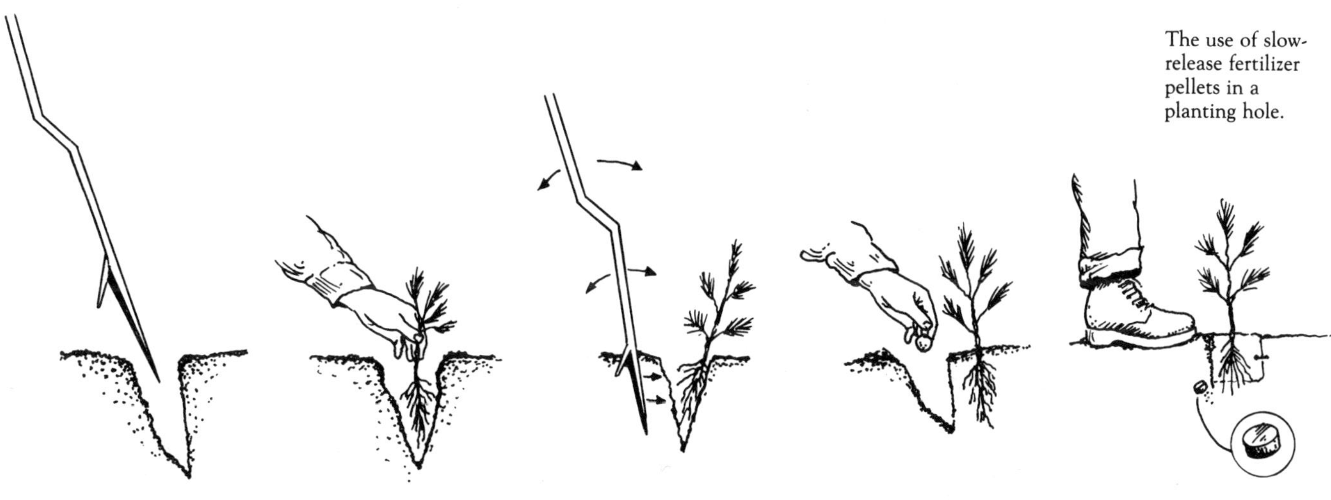

The use of slow-release fertilizer pellets in a planting hole.

holes for fertilizer placement under trees (see Chapter 44). An attachment for heavy-duty tree-planting machines has been devised for depositing slow-release materials in planting furrows.

Ring applications around each tree are generally preferable to broad-band applications between rows, although the latter is more easily done. Frequently, a pan-type spreader is used in conjunction with a garden tractor to distribute a broad 12- to 18-in. band of material between rows of trees. Pneumatic blowers are potentially useful for treating areas 50 to 100 ft on either side of access lanes in established stands.

Broadcasting and discing-in fertilizer before planting is generally undesirable because much of the applied material is not within the zone of young tree roots. Excessive weed growth, stimulated by the fertilizer, then competes for available soil moisture to the detriment of the planted stock.

Uniform distribution of fertilizer is not of major concern because the extensive and interlacing root systems, coupled with recycling of nutrients, equalize the distribution of conventional materials. The most desirable materials, from a physical standpoint, are granulated or pelleted fertilizers with uniformly sized particles that absorb water slowly. Such fertilizers, commonly available in many formulations, are particularly desirable for aerial application. Finely powdered materials drift, even under optimum conditions; and mixed goods which have a wide range of particle sizes tend to stratify in the hopper, even though continuously agitated.

A "roller" for applying anhydrous ammonia or nitrogenous gas in forests pushes the chemical through 52,000 perforation holes/acre. Calibration enables the application of 1/20 oz of N in each perforation. The piercing points are about 12 in. apart. The depth of initial penetration of the gas is about 3 in. When forest-fertilization in North America economically permits, this tool will likely be widely used.

Fertilizer applications may be delayed at least one year after new plantings are in the ground in order to reduce competition for soil, water and sunlight by the weeds which are stimulated by nutrient amendments. Losses of fertilizer by leaching and salt injury to seedlings also are minimized. Such delay cannot be justified if slowly soluble pellets or similar materials are concentrated within the root zone. Nor is the delay desirable if weeds are controlled.

Table 5.
Fertilizer Materials

Principal Nitrogen Materials

	Approximate Nitrogen Content Percent
**Ammonium nitrate	33.5
**Ammonium sulfate	20.5–21.2
Ammonium nitrate-limestone mixtures	20.5
Ammonium sulfate-nitrate	26
Monoammonium phosphate	11
Diammonium phosphate	18–20
Ammonium phosphate-sulfate	16
Ammonium phosphate-nitrate	27
**Urea	46
**Urea-formaldehyde	38
Calcium cyanamide	21
Calcium nitrate	15
Anhydrous ammonia (pressure solution for injection)	82
Aqua ammonia	16–25
Nitrogen solutions	21–49
Organic products	1–10
Magnesium ammonium phosphate	8

Principal Phosphate Materials

	Approximate Phosphate Content (Available P_2O_5) percent
**Ground rock phosphate	5*–10
Phosphoric acid	52–54
Normal superphosphate	18–20
**Triple superphosphate (concentrated)	42–50
**Monoammonium phosphate	48–53
**Diammonium phosphate	46–48
**Ammonium phosphate-nitrate	15
**Ammonium phosphate-sulfate	20
Basic slag	8–12
Colloidal phosphate*	1–4***
Magnesium ammonium phosphate	40
**Ammonium polyphosphate	60

Principal Potassium Materials

	Approximate Potassium (K_2O) Content percent
**Muriate of potash (potassium chloride)	60–62
Sulfate of potash (potassium sulfate)	48–52
Sulfate of potash-magnesia (potassium magnesium sulfate)	21–22

*Rock phosphate may contain up to 36 percent total P_2O_5

**Currently used in forestry.

***Colloidal phosphate in fine rock phosphate with clay impurities.

Weed control by cultivation or chemicals is especially important when nutrients are applied at, or shortly after, planting. In older stands, brush, vines like honeysuckle and undesirable competing trees may require control or eradication.

The influence of fertilizer applications also depends upon soil moisture. Organic soils without natural surface drainage may need to be drained in order to obtain maximum benefit from applying nitrogen and phosphate, the nutrients most often limiting growth of pine seedlings on these sites. Pine seedlings grow best where drainage maintains the water table at 4 in. below ground level, with an initial spring application of 1,000 ppa of 8-8-8 fertilizer plus 100 ppa of a trace element (micro-nutrient) mixture.

Uptake and subsequent return to the soil of nutrient elements, particularly potassium and calcium, are well-known. White birch "forages" for potassium in lower horizons and, perhaps, beyond the reach of its crown. Upon decay of fallen leaves and subsequent incorporation of foliar organic matter with mineral soil, the potassium level of the soil is appreciably improved. While white pines beyond the reach of white birch crowns grow slowly and manifest potassium-deficiency symptoms, those under the crowns of the birch are vigorous.

Similarly, eastern redcedar effectively returns calcium to the soil surface in concentrated dosages. This occurs to such a degree that earthworms, lime-loving creatures, are often abundant beneath redcedar trees and absent from the soil beyond the tree crowns. Dogwood behaves similarly in its relationship to calcium.

Vegetational influences point up the desirability of retaining certain species on sites depleted of natural fertility through agricultural mismanagement and where it is desirable to grow forest trees in the future. Thus, in severely eroded locales where littleleaf disease is widespread, growth of hardwoods and redcedar should be encouraged, regardless of the fear of undesirable hardwoods encroaching on sites where pines are commercially preferred. As calcium is deficient on littleleaf sites, the malady may be lessened by applications of lime as well as nitrogen.

Legumes and lime, alone or together, improve tree growth on impoverished upland mineral soils. Legumes (lupines, clovers, and vetches) provide a long-lasting (20 to 25 years) supply of nitrogen for sandy soils or on other sites where the influence of commercial forms of nitrogen fertilizers may last only a few years.

The following is one scheme for improving the forest with legumes:

(1) Perennial lupines are seeded between rows of trees in plantations two or three years old, when trees are no longer in danger of being overtopped by the lupine. Persisting for three to 10 years or until excluded by tree shade, the legume supplies nitrogen for the plantation and leaves, in the organic matter produced, a "reserve" of the element available after stand closure.

(2) Instead, the ground is thoroughly plowed and disced, and often lime, potash, and phosphate worked into the soil.

(3) Thereafter, the legume, properly inoculated with bacteria, is seeded in the spring and plowed under while in the flower stage during midsummer of the next year.

Liming of forests is important in improving acidic raw humus. Large quantities of nutrients are locked up in the thick layer of undecomposed organic matter characterized by this humus type. Liming greatly accelerates the biological activity that breaks down litter and encourages its incorporation with the mineral soil. Lime may aid the growth of spruce where a thick layer of raw humus exists and where legumes are grown for forest soil improvement.

The treatment improves soil structure, encourages growth of soil flora, provides an essential element and facilitates absorption of other nutrients. Overliming may cause an unfavorable calcium:potassium (Ca:K) ratio, resulting in sickly larch, Douglas-fir and Sitka spruce seedlings. Dolomite is regarded as an excellent source of lime because magnesium, often deficient for conifers, is a part of the mineral.

Lime, distributed by rotary machines or by blowers, is generally applied in the form of calcium carbonate ($CaCO_3$), but may be as the hydroxide ($Ca(OH)_2$). The hydroxide tends to become lumpy and to lose its efficiency. Applications, where needed, are 1/2 ton of the oxide (CaO) equivalent/acre for pines, 1 ton for spruce stands, and 1½ tons for hardwoods. Lime must not be applied when wet.■

Note: For documentation of literature citations for this chapter, see L.C. Walker and R.L. Beacher: Fertilizer Response with Forest Trees in North America. National Plant Food Institute, Washington, D.C.

C H A P T E R 2 3

ENERGY FORESTS

Wood was the principal source of energy fuel in the United States until 1875. Not wood that in eons of time became oil or gas or coal, but wood straight from the forest. Wood, in fact, was more important than hydropower for energy conversion until 1945.

When oil is in short supply, natural and manufactured gas is periodically rationed and coal use is curtailed because of site deterioration where it is surface-mined and air pollution where it is burned, more wood will be used for energy. (More timber will also be used for structural materials that require energy resources for their manufacture.)

Wood substitutes for other non-renewable energy resources in many ways. Here are three examples: as a source of heat, one cord produces the equivalent of almost a ton of lignite; for construction material, 6 cu ft of laminated southern pine has the supporting tensile strength of a steel I-beam weighing 180 lb, the beam requiring 243 lb of coal for production in a Bessemer furnace; and for fabrication and packaging material, minimal energy and chemicals are required (in contrast to considerable volumes of gas used to produce plastics).

In North America, fertilizer production and crop irrigation are major consumers of energy (22% and 44% of the total agricultural use, respectively, in California). About 2% of the U. S. natural gas output goes for the production of nitrogen fertilizer. Worldwide, about 5% of the gas production is used for making nutrient nitrogen.

Wood has a special potential for fuel because of its negligible sulphur content—less than 0.1%. The relatively low ash content also has value as fertilizer. To gasify wood, steam is not required and, in comparison to coal, less oxygen is needed. The gas produced, practically free of sulphur, has a higher hydrogen:carbon monoxide ratio than does gas produced from coal (called consumer gas). Facilities to produce methanol alcohol or ammonia from wood can be small operations, located near raw material in forested areas.

The forest as a living solar receptor is abundantly

Scrub oak forests, like this sandhills stand in South Carolina, could be future energy forests. More than 3,400 stems per acre, ranging from 1/2 to 6 in. in diameter, cover the land, *U.S. Forest Service photo.*

able to take the sun's rays—the only source of all of Earth's energy—and, through the photosynthetic process, produce wood. Stands of timber that are primarily managed for fuel production are called energy forests.

We refer to fossil fuels as non-renewable resources. Energy forests, on the other hand, are renewable. Carefully managed, an acre of timberland may produce as much potential energy each year as three tons of lignite. U.S. energy needs could be met by devoting only a small fraction of its land to this purpose.

Just as petroleum, gas, lignite, harder coal and peat originated as photosynthesized matter with the green chlorophyll of living plants as the chemical catalyst, so forests do the same. The big difference is that during the carboniferous era, dead and decaying plant life was suddenly buried, excluding the oxygen of the atmosphere from contact with it. As a result, total decomposition of the vegetable matter, requiring oxygen for the natural process, did not occur.

Chemical reduction replaced oxidation and the return of organic matter to the atmosphere as carbon, oxygen and hydrogen. As reduced organic matter—because of the lack of atmospheric oxygen—the carbon, hydrogen and oxygen that make up carbohydrate in the plant were bound together and locked up in the cellulose molecule. The miner and the well-driller now extract these materials from deep within the earth.

Wood for burning in fireplaces, where 80% of the heat may go up the flue, or in home furnaces is not our major concern here. Nor is interest limited to burning wood in power plants for producing electricity. The latter is practical and, in fact, wood "wastes" burned at sawmills, plywood plants and pulp mills provide over three-fourths of the industry's electric power requirements.

The present interest includes conversion of wood by hydrogenation into methanol (an alcohol) or to methane gas. Both of these destructive distillation products could be sent through pipelines now in use for transporting fossil fuels. (That trees contain readily available gas is easily demonstrated: cut into a post oak, for example, and the jet of gas squirting under pressure from the wound can be ignited and burned.)

To convert wood to natural gas, engineers use anaerobic biologic digestion, a technique observed in nature. With this procedure, chemical catalysts substitute for the anaerobic bacteria found in organic decomposition processes.

Agricultural producers of energy crops such as corn claim that cereal grains produce more calories than do tree crops. However, a greater energy requirement for the care of farm-grown vegetation is necessary. More frequent cultivation, fertilization and harvest operations than for trees make farm energy conversions significantly greater financial investments.

What kinds of woodlands are likely to be set aside for storing energy on the stump? The noncommercial scrub hardwood forests of Texas' and Oklahoma's post oak belt and cross-timbers regions are examples. Mesquite and pinyon-juniper forests and the chaparral of the Southwest are others. Commercial conifer-hardwood sites could be designated for this purpose, intensifying management so that 40-year rotations (the interval between seedling germination and final harvest) would be in order.

Sites set aside to produce trees for energy conversion are called silage plantations. Here, close-spacing (5,000 stems/acre) allows prompt and maximum site utilization. Harvesting is then carried out at perhaps five-year intervals—when the trees are less than 25 ft tall. This procedure does not require the use of heavy equipment, though a type of combine will facilitate the harvest.

For such stands, replanting is not necessary. New trees arise from sprouts, as the well-developed root systems of the severed stems, rich in starch and sugar, encourage rapid growth of young shoots. In this case, a coppice forest results (see Chapter 9).

With silage plantations, whole trees, including branches and leaves, are removed from the land and transported to conversion plants. This leads to a loss of soil fertility, as most of the nutrients removed from the land by forest trees are in the foliage. Those essential nutrients are necessary for the photosynthetic process. Hence, the ash from the conversion—and possibly commercial fertilizers—should be returned to the sites from which these fast-growing trees are cut.

Such stands, consisting of genetically superior stock, yield for American sycamore more than 16 tons of dry matter a year. River birch on certain sites outgrows the sycamore.

The value of wood for energy, contrasted to lignite, appears like this: 113 lb (3 cu ft) of oven-dry wood, burning typically at 80% efficiency, is equivalent in heat value to 1,000 cu ft of natural gas. Gas is usually burned at 78% efficiency.

Sawdust and bark burn less efficiently, decreasing as the amount of moisture increases. Consequently more weight of this "waste" is required to provide the heat equivalent of 1,000 cu ft of gas. For sawdust and bark (typically with 30% moisture content), efficiency is about 76%, requiring about 155 lb. For 100% moisture (the weight of the water equaling the weight of the wood in which it is contained), efficiency is reduced to 67%. With this much water in the sawdust and bark, 271 lb are needed to equal the energy output of 1,000 cu ft of gas.

"The worth of a thing is the price it will bring." If the price of natural gas is $3/1,000 cu ft, the energy buyer could pay up to $22.13/ton for sawdust and bark containing 100% moisture to generate the equivalent heat. Or the buyer could pay $53.12/ton for oven-dry wood. These kinds of data are conveniently read from a fuel-value calculator, a circular slide rule available without charge (in reasonable quantities) from the U.S. Forest Service.

The wood-using industry, as users and producers of energy, finds the energy crisis both a threat and a challenge. A threat because adequate techniques for burning and converting waste to energy have not been engineered to eliminate air pollution. Techniques may be feasible only at increased energy prices. On the other hand, the crisis is a challenge because it stimulates the industry to find ways to use more efficiently what is often waste material. As wood substitutes for other fuel resources, its value—and its cost—to energy industries, wood-using industries and landowners rises appreciably.

There is sufficient doomsday prophesying about population growth and consequent energy consumption without belaboring the matter. Regardless of the Federal Energy Administration's predictions of potential U.S. independence from foreign oil imports, even if the energy entrepreneurs "pulled out all the stops" to develop nuclear, geothermal, hydro, wind, ocean currents, tides, direct-solar and new fossil fuel resources, freedom from depending upon others is impractical and impossible.

Yet landowners of small forested tracts need to know that, with the exception of a few voices "crying in the wilderness," energetics people seem unimpressed with the forest as a living solar receptor. Perhaps they fail to recall that, during World War II, autos in France and Germany were powered by wood-burning furnaces in car trunks. The cars ran sluggishly—but they ran—on the methanol manufactured in those little wood burners.■

HARVESTING

The cutting of trees into logs, bolts, or poles and the skidding of these to a landing from whence they will be transported to a manufacturing plant

C H A P T E R 2 4

LOGGING THE FOREST—BEST MANAGEMENT PRACTICES

"EPA has developed a system of guidance for evaluating non-point sources (of pollution), including silvicultural activities. The system is guidance; it is not regulation." So reads a statement from the Environmental Protection Agency (EPA). That statement would calm the fear of many, were it not for the notoriously poor record of federal agencies which set out to guide and then, having felt the pulse of power, proceed to regulate.

For this discussion, we assume "guidance" to be a continuing attitude of EPA and, hence, try to outline procedures for checking unnecessary non-point sources of pollution of the nation's waterways attributed to the management of small forested tracts. These procedures, to use EPA terminology, are best management practices (BMPs).

Federal legislation enacted in what are commonly called amendments to the Federal Water Pollution Control Act, contained in Public Law (PL) 84-660, "encourages cooperative activities by the states for the prevention and control of water pollution." This legislation also enables interstate agreements to encourage passage of relatively uniform laws by adjoining states.

Subsequent legislation (PL 92-500) requires the EPA administrator to issue "(1) guidelines for identifying and evaluating the nature and extent of non-point sources of pollutants, and (2) processes, procedures, and methods to control pollution resulting from . . . silvicultural activities, including runoff from . . . forest lands" (Section 304). Section 208 of this Act, entitled "Areawide Waste Treatment Management," calls for a plan to "identify, *if* appropriate . . . silviculturally related non-point sources of pollution, ..." The law is unambiguous in its intent to include the country's woodlands in EPA's jurisdiction, though the words emphasized leave room for argument as to how definitive the regulations of the agency may ultimately be.

The forest landowner, as the steward of at least a portion of a watershed, has some intrinsic interest in maximum production of clear, clean water. As timber manager, the landowner aims to grow maximum amounts of fiber. In managing lands for both water and wood, the goal is to optimize, by compromise, the amounts of both of these renewable natural resources generated.

As a rule of thumb, a fully stocked forest consumes the precipitation equivalent of 17 in. per year. Cutting all of the trees maximizes water production. If regrowth is prevented (as by chemical herbicide treatment or bush-hogging), timber production is obviously minimal: it is zero.

Removing the severed trees from the area in this example would constitute clearcutting. That harvest, coupled with other silvicultural treatments, would maximize fiber production. Available water, too, would be increased by the 17-in. equivalent;

Simple Conshocton wheels measure water runoff and sediment carried from an area by water as a result of various logging practices. A 1/100-volume sample passes through the slot as the wheel spins on precision ball bearings in the axle.

but now the water has a higher sediment load.

An optimum condition could be attained by cutting only a portion of the trees (thus lessening the amount of increased out-flow of water) and making the cut in such a way as to permit only minor amounts of soil to be washed from the harvest site. Cutting in strips of 2-chain (132-ft) widths on contours is one such technique. Selection harvests are another.

At this point, we note the reasons these two kinds of regeneration harvests are used. Clearcutting—whether in strips, blocks, or patches—could appropriately be called an ecological harvest method. It is a necessary technique for certain species which require the mineral seedbed to be exposed for receiving seeds and for their germination and full sunlight for growth of seedlings, saplings and (sometimes) pole-size stems. Commercially marketed species subject to these ecologic constraints include the southern pines, ponderosa pine and Douglas-fir.

On the other hand, selection harvests—taking perhaps 10% of the stand at 10-year intervals—are appropriate for shade-tolerant, heavy-seeded hardwoods like the oaks and hickories of the South and the spruce and true firs in the higher elevations of the Rocky Mountains and Far West.

In between these two harvest extremes are the seed-tree and shelterwood methods. Like clearcutting, they result in even-aged stands. For a seed-tree harvest, 5 to 10 trees/acre are left for a seed source. For the shelterwood system, the harvest is usually made in three cuts, five to 10 years apart, some trees being left to provide protection for the seedlings that germinate shortly after the first cutting. Seed-tree and shelterwood harvests aim to do what clearcutting does—regenerate an even-aged forest of a single species. However, the cost for the former is much greater. Loggers need to return to the site to harvest stems not taken in the first cut, while lightning and insects kill some trees left for the later harvests. Additions to the cost of harvest are usually paid by the consumer—the buyer of a house and the reader of the daily news. Sometimes timber growers absorb these costs to protect their investment; sometimes they are not able to pass on these increased expenses.

BMPs may vary with land ownership. The federal government, even though its budget limits what can be spent for water-quality protection, is not excused from using BMPs on national forests. The small, private landowner also is not excused in spite of his need to make his profit at the stump. For this individual, however, the Section 208-required plan must allow owners a profit that will encourage them to continue to manage the forest. The law is clear that there will be one level of BMP to encourage landowners (whether large corporations or small farm owners) to be motivated to maintain clear, clean water.

Violation of the Federal Water Pollution Control Act may subject a landowner to fines of $2,500 to $25,000 per day. Many of the so-called violations are for standard practices, generally considered routine to a landowner's operations.

Silvicultural operations are not likely to be concerned with point-source pollution, alluded to earlier. Point-source pollution for forestry practices is limited to log sorting and log storage, using water that is subsequently discharged into streams classified as navigable waters. Rock crushing and gravel washing of road-bed material is also classified as an activity subject to point-source pollution restraints.

Non-point sources of concern are principally those resulting from sedimentation of streams, and fertilizer, herbicide and insecticide leaching into underground aquifers. Those chemicals eventually percolate into surface waters via springs or directly contaminate creeks, ponds and lakes. Fungicide and natural salt leaching, the latter from decomposing rock gravel used on logging roads, is theoretically possible, but improbable in forestry. Sedimentation and salt leaching may follow prescribed burning.

BMPs designed to restore and maintain the chemical, physical, and biological integrity of the nation's waters include many long-used techniques. Among them are

1. building waterbars,
2. putting unused roads and trails "to bed,"
3. promptly regenerating harvested areas,
4. controlling road grades,
5. building roads on contours,
6. contour harvesting,
7. skidding up-grade to landings,
8. constructing culverts properly,
9. placing roads away from streams,
10. removing logging debris from stream channels,
11. restricting streamside harvests,
12. prescribing fire properly and

13. using chemicals judiciously.

A few lines about each:

1. *Waterbars:* Waterbars may be slight mounds of soil running diagonally across the trail or a trough made of 2x8-in. wood planks, inserted flush with the road surface. Spacing of these impediments to surface run-off on logging roads and skid trails will depend upon grades. On roads, bars should be spaced as close as 125 ft on 15% grades and may be as far apart as 300 ft on grades of 2%. For skid trails, the range runs from 25-ft spacing on 45% slopes to 250-ft spacing on 2% inclines. Waterbars should be maintained daily during logging so that they neither fill with sediment and debris nor are allowed to be destroyed by heavy trucks and skidders.

2. *Putting Roads and Trails to Bed:* Roads and trails no longer needed for logging may require seeding, fertilizing, and liming of the soil. Grasses and legumes are effective stabilizers. Game-food vegetation may be planted.

A recommended general practice is to:

(a) prepare the site with a chain-type spike-hook harrow attached to a crawler tractor,

(b) apply 1/2 ton/acre of a complete fertilizer, such as a 2-12-12 N-P-K formulation,

(c) apply 1 ton/acre of limestone to supply adequate calcium and to raise the pH, and

(d) plant 20 lb/acre of a high-quality seed mixture for wildlife forage, such as shade fescue, hard fescue, Kentucky bluegrass and meadow fescue.

Another recommendation, following roadbed preparation, is to:

(a) apply 2 tons/acre of limestone,

(b) fertilize with 1 ton/acre of 2-12-12 plus 1 ton/acre of 20% superphosphate and

(c) plant orchard grass and Ladino clover in the spring at rates of 13 and 3 lb/acre, respectively.

Lespedeza, rye, and other cover crops are also employed.

Soil tests enable more precise prescription of fertilizer needs.

3. *Prompt Regeneration:* A new stand should be assured within a year of harvest. If a seed source is not available, or drought destroys a new crop of natural seedlings, planting of nursery-produced stock or direct-seeding (as a farmer sows wheat) are appropriate practices. Delay is more apt to result in weeds destroying the seedbed than in excessive run-off of water and soil.

Any cover on the soil will reduce the impact of rain. A 2-in. layer of pine straw holds as much as 0.1 in. of water after a rain, thereby reducing erosion by providing time for gradual infiltration of precipitation into the surface of the soil. There are about 20,000 lb/acre of litter under pines in old fields when the weight has reached equilibrium at an age of 20 to 35 years. Conifer litter is several years behind broadleaf trees in accumulating because foliage persists on needleleaf trees. (Baldcypress, larch, and tamarak are exceptions.)

Natural leaching of minerals increases following clearcutting. The losses into streams represent that portion of the minerals in the soil ordinarily absorbed by the trees and annually recycled through foliage decomposition. Shortly after new vegetation covers the land, the amount of salt discharge returns to normal. Normal is not zero: if it were, how would the oceans have become salty?

4. *Controlling Road Grades:* Grades should rarely exceed 8%. Lesser grades, on the other hand, mean longer roads and, therefore, a larger area from which roadbed and roadside erosion may occur. Too, the more area dedicated to roads, the less on which to grow trees.

5. *Contour Road Layouts:* As with contour terracing and contour plowing, building roads as nearly on the contours as possible significantly reduces erosion, both during construction and thereafter. Roadside banks must be seeded as soon as the grade is established.

6. *Contour Harvesting:* Removing trees in contour strips and skidding logs from stumps to landings (where logs are assembled for loading on trucks) along the contours is encouraged.

7. *Skidding Up-grade:* Each log dragged across the ground gouges out an erosion channel in its skid trail. If all logs are pulled down-slope to a landing, the landing will be as a palm of a hand with many fingers, the fingers being the skid trails. Water running down these finger channels accumulates at the palm, washing out the road where the landing is located. With the hand turned so that the palm is up and the fingers point downward, water originating at the landing is dispersed as it flows from the landing into the many finger-like channels made in the ground by the skidded logs.

8. *Culvert Construction:* This, like the placement of slight depressions in the hairpin curves of logging

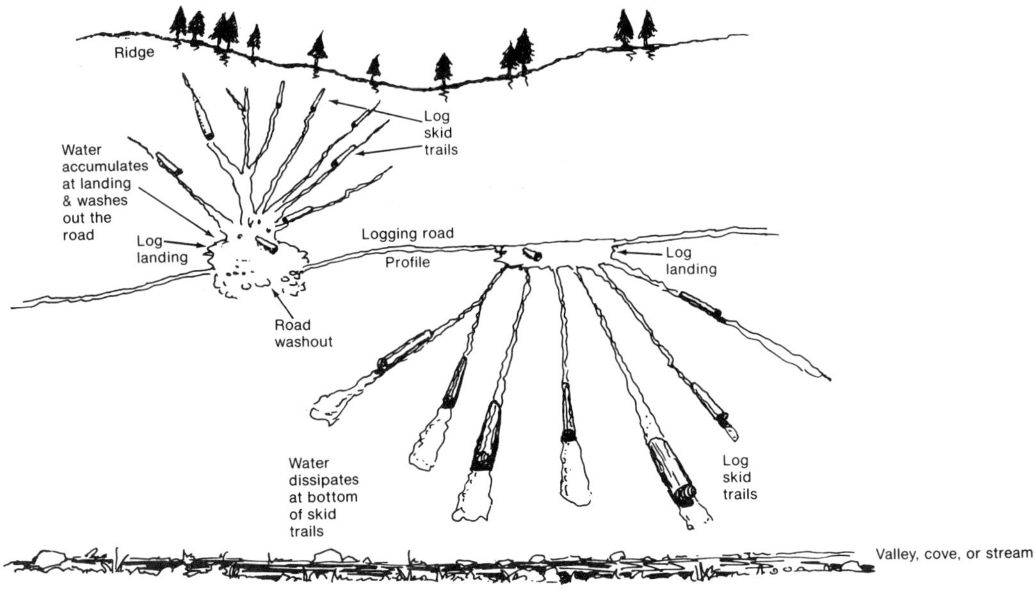

How skidding up- and down-slope influences erosion.

roads where intermittent streams are crossed, requires skill and large equipment. Improperly placed structures wash out, causing serious erosion.

9. *Avoid Placing Logging Roads Near Streams:* This rule of thumb serves for suggesting the minimum distance between a logging road and a stream where some sediment is acceptable. Starting with a strip 25 ft wide on level land, the width (distance from stream) may be increased 2 ft for each 1% increase in slope between road and stream. For watersheds where no appreciable sediment is permitted, logging may start 50 ft from a stream and the distance increased 4 ft for each 1% increase in slope. Culvert spacing, road surface conditions, steepness of road grade, amount of sediment trapped in small coves and amount and arrangement of logging slash and fallen tree debris (whether impeding or dispersing water) are factors which modify the minimum acceptable distance between roads and streams.

10. *Removing Debris From Streams:* Under some regulations, laborers walk all stream courses, both permanent and intermittent, removing logging debris that has found its way to these sites. Such material may obstruct the natural flow of water as well as contaminate streams with organic impurities.

11. *Stream-Side Harvest Restrictions:* Harvest of trees no closer than 1/2 chain (33 ft) on either side of a stream is encouraged to reduce sedimentation of the water course. As a result, for every mile of stream, eight acres are removed from commercial timber growing.

12. *Prescribing Fire Properly:* Prescribed fire need not cause significant soil movement. Prescriptions can be written which should satisfy erosion limitations. For instance, a late afternoon winter fire, running with the wind when the relative humidity is about 30%, the temperature about 50°F, and ignited three days after a 1/2-in. rain will not likely cause soil movement.

Destruction of humus exposes soil to the impact of raindrops. Where humus layers are burned, exposure of the soil may be sufficient to seal pores and reduce infiltration to a negligible rate. Continual burning retards the formation of a desirable humus layer by eliminating litter that would in time become incorporated into the mineral soil. Fire may also significantly reduce populations of organic matter-decomposing fauna and flora living in the unincorporated litter of the forest floor. Although occasional wildfires in pine stands reduce growth, probably due to defoliation and cambial injury rather than to soil destruction, annual burning under severe conditions causes deterioration of soil water-holding properties.

13. *Chemical Leaching:* Fertilizers, insecticides and herbicides are sporadically used in silvicultural operations. Even though applied to the land or its vegetation at 20- to 50-year intervals, monitoring systems may need to be devised.

Costs usually preclude the use of fungicides in American forestry practices. Therefore they need not be a concern for BMPs.■

C H A P T E R 2 5

LOGGING THE FOREST—THE METHODS

This chapter does not intend to give the landowner the ability to conduct a logging operation. The procedure is too risky for the inexperienced. Rather, the purpose is to enable the landowner to understand the why and how of logging operations, at best a difficult and dangerous task. Moving 1 ton in a single not-so-large log is no easy feat.

Logging involves

1. constructing a road to the timber,
2. clearing an access to the tree to be cut, sometimes providing (with the aid of a bulldozer) a bed of brush, logging slash, or loose soil on which the tree is to fall,
3. felling the tree,
4. limbing (really delimbing) its branches,
5. bucking the bole into log lengths (from 8 to 36 ft, depending upon products to be made, equipment for the task, and economic considerations),
6. bunching a group of logs together in the woods,
7. skidding them to a log "landing" or "set" (called yarding), and then
8. hoisting or cross-hauling the logs onto trucks for transport to the mill.

Felling on small forested parcels will usually be by power chainsaws. (Larger industrial operations may use massive shears.) Under some circumstances, such as where the noise of a power saw is forbidden, two-man crosscut saws come in handy. The more mechanized the equipment, the lower will be the stump. Those who have labored with the crosscut readily understand why stumps of trees cut with those back-breaking tools were often waist high.

Limbing is done with power saws and axes to a merchantable top. At that point, the top of the tree is severed. It remains in the woods.

Bucking cuts the bole into log lengths convenient to handle: 2-ft bolts for roof shakes, 4- or 5-ft bolts for pulpwood, 8-ft logs for crossties, 17 ft for sawmills, 35-ft timbers for pilings or for some sawmills, or tree-length (to the highest point on the tree considered merchantable, perhaps exceeding 50 ft).

Woods-workers should wear hard hats and, when using power saws, leg protectors. By not using safety equipment, this man risks injury.

Measured marks must be made on the boles, utilizing a measuring stick, prior to bucking. The logger makes adjustments in log lengths to maximize the volume of high-quality lumber that can be eventually sawn from the tree.

Economic success of the lumbering operation depends upon how well these practices are carried out. A tree shattered because it missed the felling bed, logs bucked too short or too long that waste wood and branch stubs that interfere with the otherwise-smooth operation of the sawmill carriage cost the buyer money.

Transporting logs to landings will likely be by rubber-tired skidders for small timberland sales. Small track-type tractors may also be used, although the tractor will compact and/or puddle the soil, thereby reducing its future productivity. Mule- or horse-logging is again becoming important in American forestry as landowners often are willing to accept less money for their stumpage in return for timber removal that does minimal damage to residual trees and to the soil. Draft animals provide this incentive. (Industrial loggers use helicopters, balloons and cable systems to move logs from where trees are severed from their stumps to landings.)

Chokers placed on logs bind the cable to the log so that it will not come unhooked during the skidding operation. Single logs may be bunched together in the woods in groups for subsequent transport to the landing. This saves money, wear on equipment or animals and damage to the soil. The bunches are then picked up and hauled to the landing by forwarders—tractors equipped with grapples or loading booms—or by animals.

Cross-haul loading, referred to earlier, involves the use of animals or small tractors to roll or pull logs up poles to the beds of trucks. Stationary and portable cables and tractor-powered tongues load the logs at the landing on to truck beds. Trucks used in the woods occasionally pull loaded log trailers to a public road, where the trailer is unhooked and attached to a cab that satisfies state safety requirements for the remainder of the trip to the mill.

Pulpwood and fence posts are treated much as sawlogs and other large timbers through the bunching process, although pulpwood cutters use bowsaws, rather than crosscut saws, if chainsaws are not available. Usually pulpwood harvests take place in stands in which trucks can be driven to the felled tree: loading equipment is either not necessary or the size requirements for the rigs greatly reduced.

In most areas, fellers also buck the trees they have cut. Other people working in the logging operation may be bunchers, skidders, and loaders. In small-quantity sales, the same workers may do all the tasks, laying out the plan of work for safety; skidders avoid areas in which trees are being felled. In some areas, felling and skidding are not done on the same day. High in the Rocky Mountains, felling and bucking may be done in one summer, skidding and loading in the next.

Landowners should express concern about roads and skid trails on their property. Grades seldom need to be more than 8%. Specifications can be included in the contract. Steep grades may seem to enable cheaper logging than more modest ones, but not if the logger has to use his largest tractor to push loaded trucks up grades during wet weather. Steeper grades also require more braking power in descent, more fuel for climbing and offer a greater chance for trucks carrying loads of logs to run away.

The principal reason for gradual grades is to protect the road from washouts and to protect the logging site from erosion (see Chapter 24). Gully-washing, trash-moving rains destroy steeper-graded roads, as water cascades down a roads' lower side to create gullies in the forest.

To protect the road and site, skidding to landings should be upslope, thus when water concentrates at the landing, it dissipates down the many fingers of skid trails. If skidding drags logs downslope to a landing, the water from all of the scarred log trenches accumulates at the landing and washes it out. A cutting pattern that follows contours may be a happy compromise.

Landowners may control, by a statement in the contract, the weather conditions under which logging is permitted. The possibly reduced price for stumpage could be more than returned by maintaining the future productivity of the site. Loggers, of course, want to be free to work in the woods whenever the weather permits them to do so, not when damage to the site will be minimized. They hold back for such times readily accessible tracts, those on drier sites, flat lands, and near to paved roads. A contract item could read thus: "Logging may be done only after three days have lapsed since a rain of ½ in. or more." (Because of fire danger, similar statements, but at the other end of the precipitation scale, may control the periods when loggers are permitted to work on an owner's property.) ■

ADMINISTRATION

Management of the forest apart from silviculture

C H A P T E R 2 6

ESTIMATING LAND PRODUCTIVITY

The productive potential for land to grow trees depends upon physical and chemical properties of the soil, climate, lay of the land and living organisms that may encourage or discourage the vigor of the particular species desired for timber. Site index (SI) measures the ability for land to produce wood. SI denotes the average total height of dominant and codominant trees at age 50. For example, using the loblolly pine curves accompanying this chapter, dominant and codominant trees that are 87 ft tall at age 50 are growing on SI 87 land. The same trees on the same land at age 40 are 69 ft tall; at age 60, 94 ft.

Site indexes are also designated for other ages, but tables and curves must so indicate if that is the case. For Douglas-fir forests in the Pacific Northwest, the site may be indexed for trees at 100 years of age. The figure then is for SI-100. Otherwise, age 50 is assumed. Original SI curves and tables give values for unmanaged, fully stocked, even-aged stands of a single species. Measurements of trees under other conditions or the selection of intermediate or suppressed sample trees drastically skews SI. Later-developed data provide for managed, second-growth forests and plantations. Curves like those shown are used to determine SI for trees of any age.

To determine the productivity of abandoned land or of a recently clearcut forest where no trees are available for age and height measurements, soil characteristics that relate to tree growth are noted. For instance, in the West Gulf Coastal Plain, loblolly pine growth depends on the texture of the subsoil, depth to the subsoil and the surface drainage of the land. Generally, the finer the subsoil, the more rapid the growth. Growth is also better with deeper topsoil and slower surface drainage. SI, hence, may vary from about 70 to more than 100, depending

Site index (SI) curves like these are available for many species. SI measures the height to which dominant and codominant trees are expected to grow in 50 years; SI-100 denotes age 100; and SI-25, age 25. Natural stands of eastern white pine (a); planted stands of eastern white pine (b); southern cottonwood (c); American sycamore (d); upland oaks (e); sweetgum (f); yellow-poplar (g); shortleaf pine (h); longleaf pine (i); natural slash pine (j); planted slash pine (k); natural loblolly pine (l); planted loblolly pine (m); lodgepole pine (SI-100) (n); Douglas-fir) (o,p) [k & m, Lenhart *et al.*; n, Briskell; o, Carmean; p, others, miscellaneous sources].

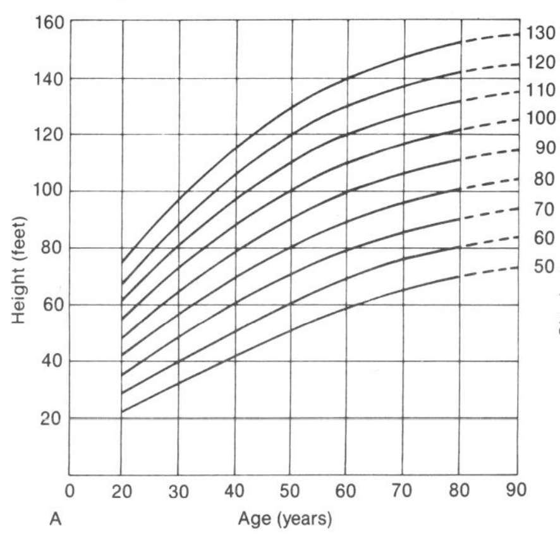

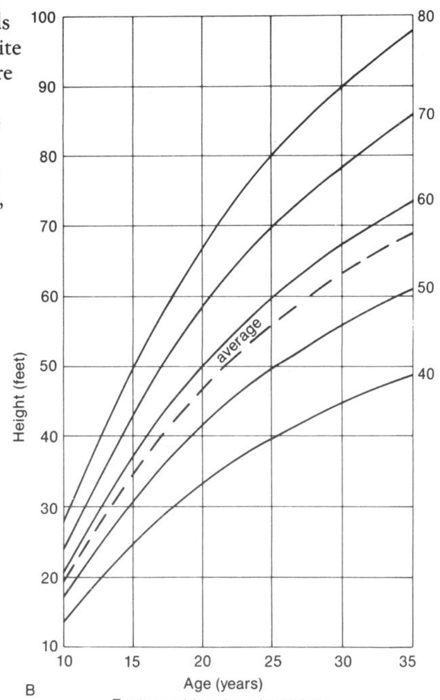

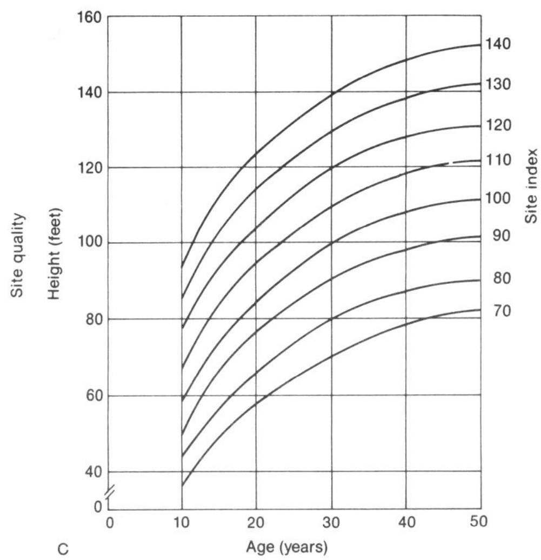

upon these three characteristics.

For slash pine, widely planted throughout the South, growth rate is also related to drainage and to soil texture. Good water movement gives better growth on shallow soils. The opposite is true for deeper soils.

Longleaf pine productivity in the West Gulf area on sites that are well-drained also depends upon subsoil texture. Here, the finer-textured silt and clay soils provide for faster growth than does coarse sandy soil. Except for this species, SI ordinarily is not affected by the number of trees per acre (stocking) or the rate of diameter growth. Dense stands usually indicate higher-quality land. Where longleaf pines have sufficient soil water and nutrients for rapid growth, it is because the fine soil particles hold moisture and nutritional elements.

Longleaf pine, in the lowlands along the Gulf coast, is found on poorly drained sites. There it often grows faster than in its more typical habitat on high, dry ground. It is not unusual for longleaf pine, once past its grass stage—a dwarf condition lasting from several years to more than a decade—to overtake slash pine in 20 years or so.

Depth to the mottling zone in poorly and imperfectly drained soils may also influence growth for longleaf pine. If the variegated coloring—caused by oxidation, hydration and reduction of the iron coatings on soil particles—occurs as shallow as 18 in. in a heavy clay zone, SI will likely be about 70. If as deep as 4 ft, SI will be 10 points higher.

Table 6. Soil Sites

Site index of longleaf pine on poorly- and imperfectly-drained soils in the southeastern Coastal Plain as influenced by soil characteristics (from Coile).

Subsoil Characteristics	Depth to Mottling Inches		
Texture	18	30	48
		Site Index	
Sand	59	62	67
Loamy sand	60	63	69
Sandy loam	62	66	71
Sandy clay loam	64	68	74
Sandy clay	67	70	76
Light clay	69	72	78
Heavy clay	71	74	81

Coastal Plain bottomlands have a somewhat greater SI than adjacent uplands for many species. Heavier-textured soils are typical in most of these bottomlands. Increased growth is due to the enriched mineral matter and humus leached or washed from higher ground. In this deeper soil, moisture is more available during drought periods than in the shallower upland soils.

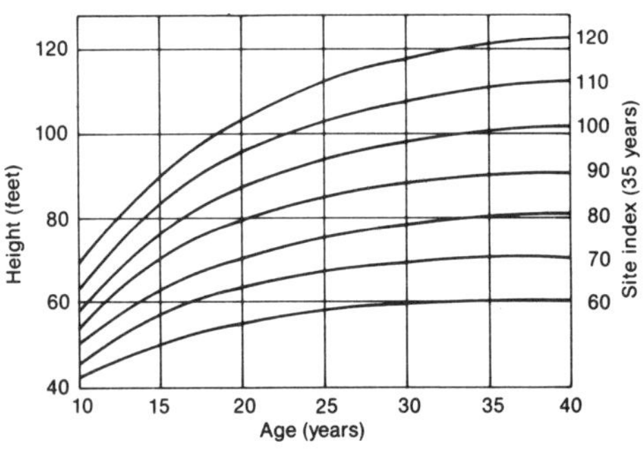

D American sycamore

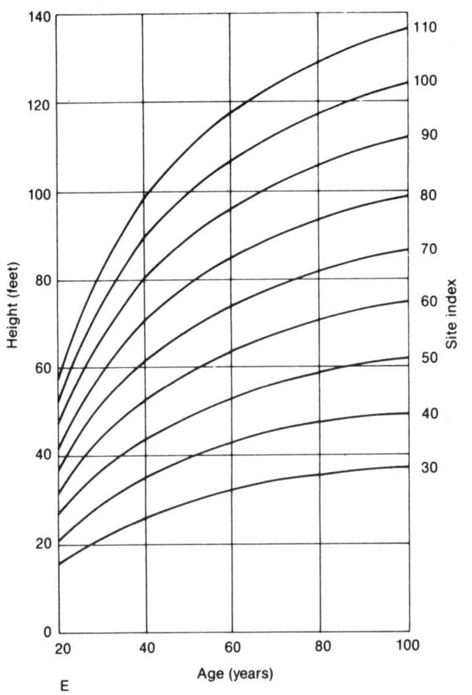

E Upland oaks

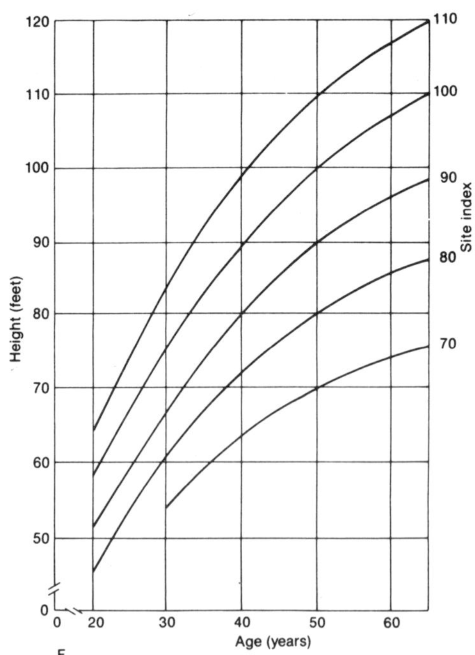

F Sweetgum

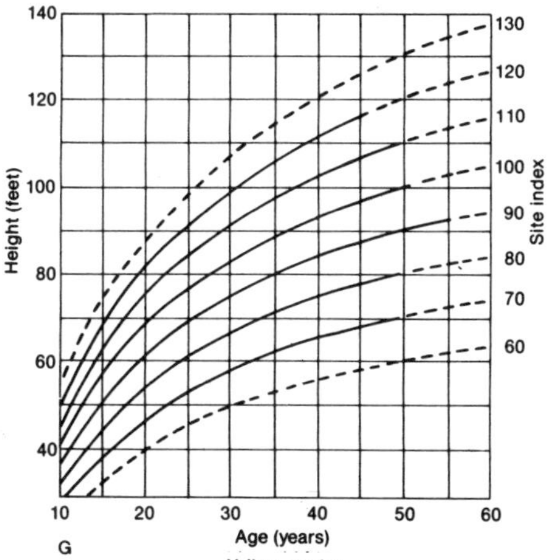

G
Yellow-poplar

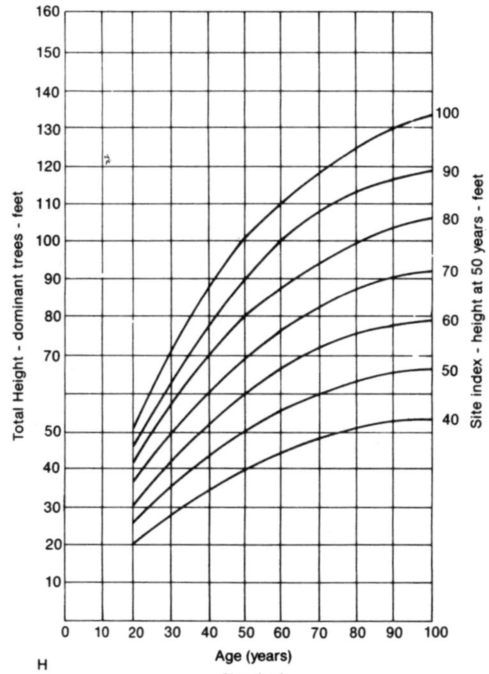

H
Shortleaf

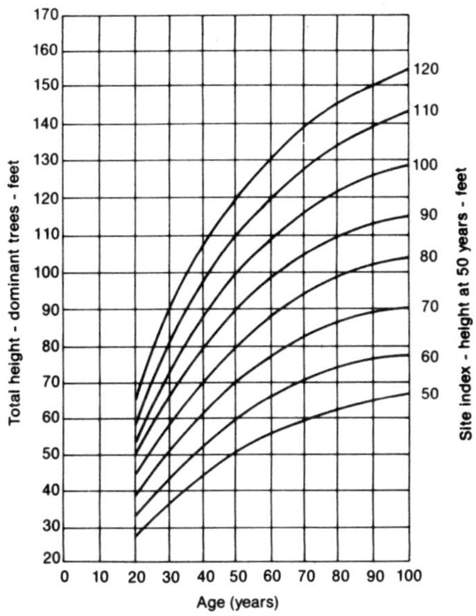

Longleaf

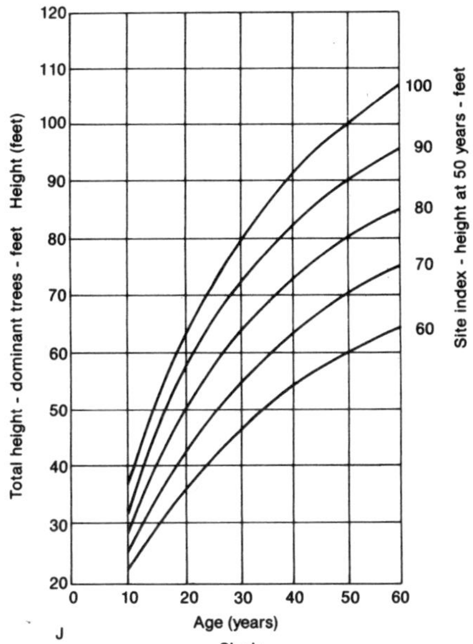

J
Slash

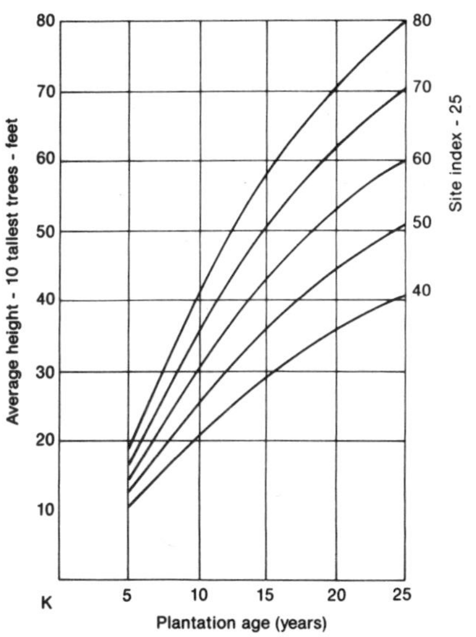

K
Slash pine
Plantations

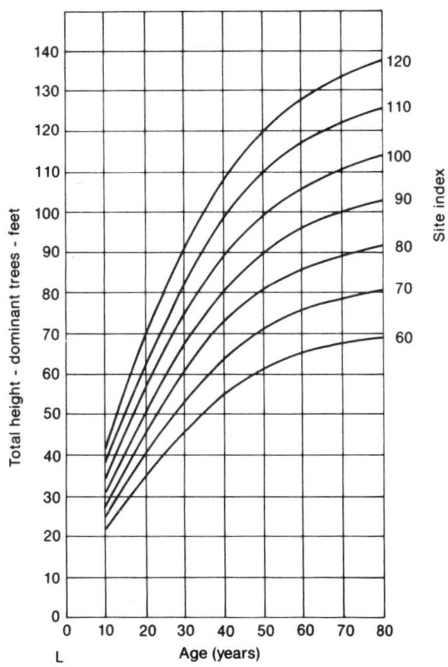

L
Loblolly pine

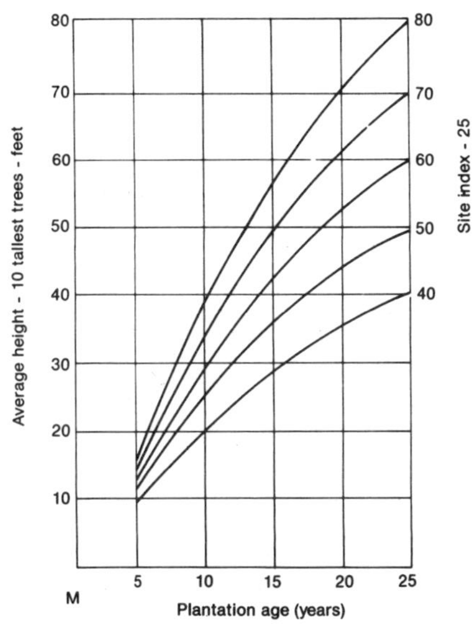

M
Loblolly pine
Plantations

Bottomlands and creek bottoms are the ideal habitat for many hardwood species in the eastern and southern U.S. Growth of individual stems selected as crop trees will be rapid. For this part of the country, reliable techniques for estimating growth of the dozen or more commercial hardwood species found in stream bottoms have not been developed.

Coastal Plain soils are relatively variable. Deep sands, often interspersed with shallow soils, may be underlain by "pans." These hard, dense, clay layers prevent water and root penetration both up and down. Hardpans are caused by the "rusting" together of the iron coatings on the surfaces of soil particles. They may be as little as 1 ft below the surface, but many feet thick.

Hardpans greatly affect tree growth by controlling water availability. During droughts, available water may be "locked" below the pan, unable to rise. Conversely, for days after a rainy season, when deep soils have dried out, the shallow surface soil above a pan may still be moist because water has been trapped in a "perched" water table. Growth is poor on such sites. Subsoiling—cutting through the surface and subsurface layers of hard ground—improves drainage and thereby increases survival and growth. Subsoiling is usually done with a knife-like attachment on a farm tractor or small dozer.

As an indicator of tree-growth rates and of preferred species composition, soil pH is a useful tool. All southern pines, for example, grow best in soils with an acid "reaction," about 5.5 on the pH scale (where 7 is neutral). Generally, the best sites for broadleaf trees will be neutral or only slightly sour, perhaps 6 to 6.5. The junipers, commonly called cedars, while found growing in many sites, prefer limestone-derived soils. They are noted for good growth in alkaline lands with pH measurements approaching or exceeding 8.

As a generality, lower lands are best for forest-tree growth and higher elevations the poorest. One notes this by observing the tops of trees on undulating lands: tops are level although the land is gently rolling.

In some regions, certain vegetation called indicator plants suggest which species are better to retain on the land and the rate of growth of recommended kinds of trees. Many of these relationships have been charted in ecographs.

Landowners need an idea of tree-growth potential in order to determine on which sites to spend money for cultural practices, including planting, weed-tree control, thinning and regeneration harvests. The difference between a 10- and a 20-in. depth of surface soil over a poorly drained subsoil could mean a difference of 30 ft in height for a stand 25 years old. That could translate to a difference—depending upon scale and conversion factor—of over 20 cords, or 10,000 bd ft, per acre. The difference

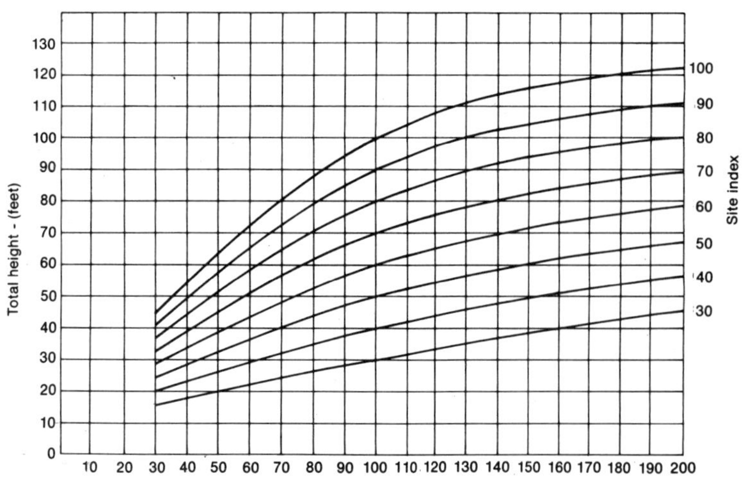

N

O

Average Douglas fir (SI 140) curve

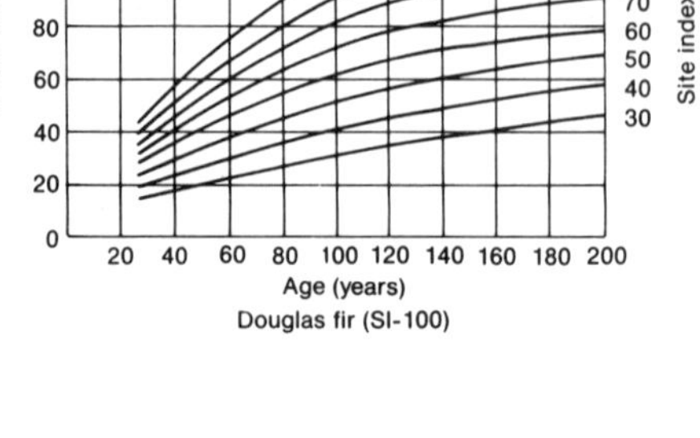

Douglas fir (SI-100)

in value attributable to site quality, apart from the use of any management procedures, would be appreciable.

While some landowners may prefer to put all grades of land—even their poorest—into production, others prefer to put their money and effort on the highest-quality sites, allowing less-productive land to lie idle. Fertilization, for instance, may give the most dramatic improvement in growth on infertile sites. The cost of that treatment, on the other hand, may be more profitably used to rid stands on high-quality land of worthless weed trees.

The Soil Conservation Service and some forestry organizations in the West use five classes for describing site productivity. Site I typically occurs on alluvial benches alongside streams; Site III is intermediate; and Site V the lower-quality land of shallow or droughty soils.■

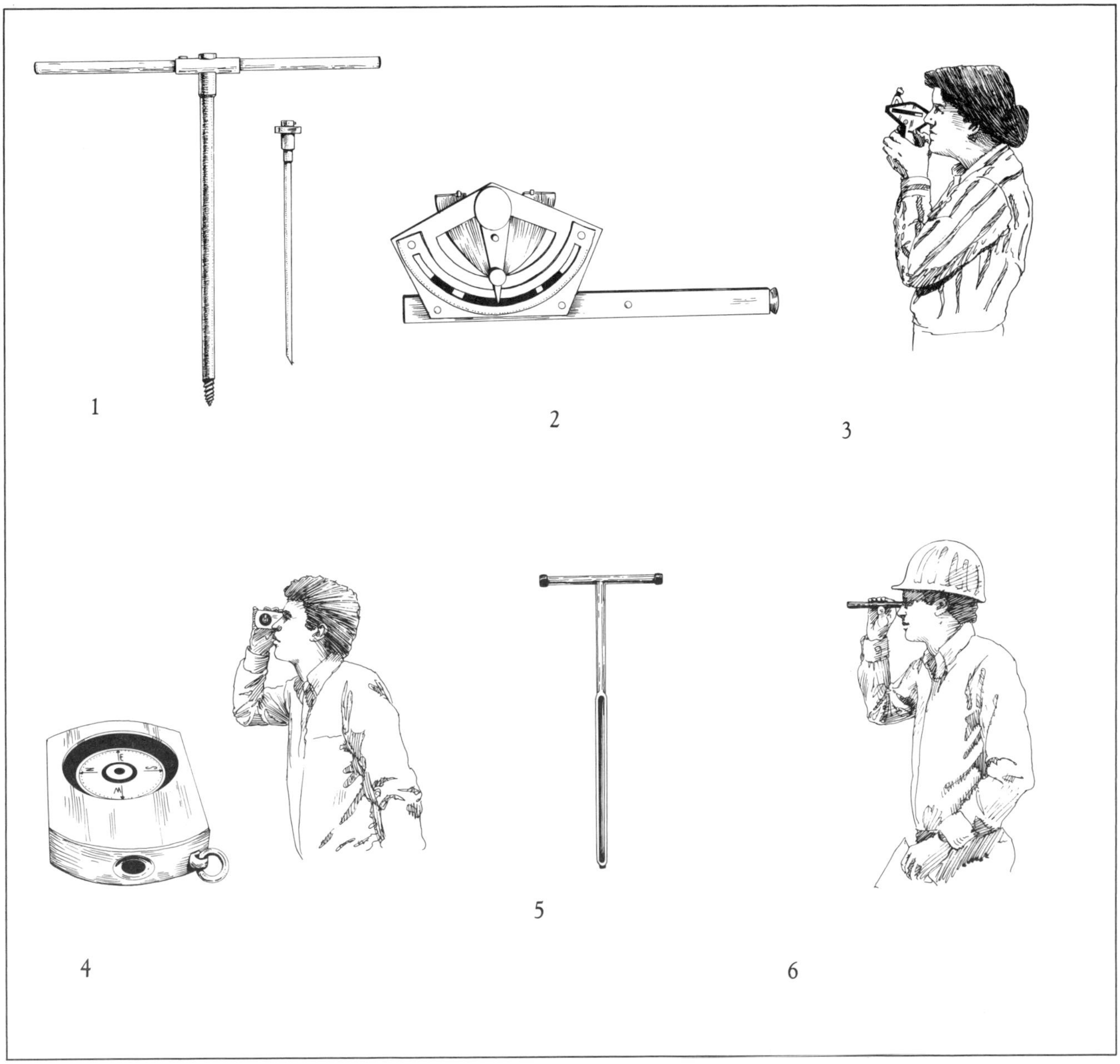

Tools of the forest manager: (1) increment borer and extractor, (2) Abney hand level for height determination, (3) Haga altimeter for height determination, (4) combination height finder and compass, (5) soil sampling tube, and (6) an angle gauge for determining basal area.

CHAPTER 27

MEASURING TREES AND LOGS

Landowners and timber buyers sell and purchase wood by tree scale, log scale and weight. For tree and log scales, measurement is in board ft, or thousand board feet or thousand board measure (MBF or MBM). Weight tallies are in tons. A board foot is a piece of wood 12 in. x 12 in. and 1 in. thick, or the equivalent. In practice, because of saw kerf—the width and wobble of the blade—7 to 8 bd ft are cut from 1 cu ft of wood. As solid wood, 1 cu ft is literally 12 bd ft.

Log scales, the means by which originally all American wood sold was measured, are classified as *diagram* or *formula*. J.M. Scribner, a preacher in the mid-1800s in the northeastern logging country, used cross-section paper to figure out how to cut up logs in order to maximize a sawyer's production at the mill. From his graphs, he could determine the number of 2x6s, 2x8s, and 1x10s, for instance, that could be obtained. Thus, a log 16 ft long and 14 in. in diameter inside the bark (i.b.) at the small end can yield 110 bd ft; one 21 in. i.b. at the small end of a 20-ft log scales 380 bd ft.

Diameters are measured 4½ feet above the ground.

At about the same time, Edward Doyle, versed in mathematics—in his day and by the forester still called mensuration—used a formula to obtain the volume that could be cut from a log. Thus $V=(D-4)^2$ gives the volume of a 16-ft log. For logs of other lengths,

$$V = [(D-4)/4]^2 L$$

where V = volume in bd ft,
D = diameter inside the bark at the small end of the log, and
L = length of the log.

Doyle deducted 25% from the solid wood content to allow for the saw kerf. The -4 in the formula takes care of slabbing, the waste that results in sawing the log.

Scribner's rule generally scales high for logs less than 18 in. in diameter, possibly giving the buyer less wood than he pays for. On the other hand, Doyle's rule scales high for logs larger than about 24 in. possibly giving the buyer less wood than he pays for.

Another popular rule, the International, is also a mathematical scale. It was developed from the equation: $V = .22\ D^2 - .71\ D$, V being the net board-foot volume in 4-ft cylinders and D the diameter of the cylinder at the small end. As an example, for 16-ft logs, four 4-ft cylinder volumes are added together in the scale volume table.

Landowners may come upon buyers who use another, sometimes a company, scale. An owner may sell by one scale and the buyer purchase by another.

Subsequently, this log scale information was put on pieces of wood called scale sticks. The scaler determines the length of the log and, with the stick placed over the cross-section of the small end of the log, reads directly the volume in board feet. These sticks, available from supply houses, are inexpensive.

By the 1940s, when sellers were merchandizing their trees while the stems still stood in the forest, a more accurate procedure than the timber estimates of buyers became necessary. Thus were born tree-measurement scales adapted from the log scales.

A typical and convenient scale for the small woodland owner is the Scribner decimal C. With this measure, volumes are rounded off to the nearest 10 ft. Thus, a 12-in., two-log tree is scaled at 8 instead of 75 and a 12-in., three-log tree tallies 10 instead of 95. After all addition and deduction for defect is done, a zero is added to the total. The decimal C scale is adequately accurate for the timber in sales made on relatively small parcels of land.

Trees must be measured at breast height (4½ ft above the average ground line) when using tree-measurement rules. The diameter tape (D-tape) must be pulled taught and be absolutely perpendicular to the bole or error which favors the seller will creep into the measurements. D-tapes are inexpensive, more accurate, and less cumbersome than calipers and Biltmore sticks, also used for this purpose. (Calipered measurements from several angles

Partial Field Tally (a), Stand Table (b), Scribner Dec. C FC 76 Table (c), and Stock Table (d)

d.b.h.		Number of 16-foot logs							TOTALS
		1	1½	2	2½	3	3½	4	
(a)	10	⊠ :	⊠ ·	⁖					
	12	⊠⊠ ·	⊠⊠⊠ :	⊠⊠ l :	⊠⊠ L :	⊠			
	14	··	⁖	⊠ ::	⊠⊠ ·	⊠	··		
	36				⊔	⁖	:	·	
(b)	10	12	11	3					26
(a) = (b)	12	21	32	25	26	9			113
	14	2	3	14	21	9	2		51
	36				1	3	2	1	13
TOTALS		35	46	42	54	21	4	1	203
(c)	10	3	3	4					
	12	4	6	7	8	9			
	14	6	8	10	12	14	14		
	36				115	134	148	163	
(d)	10	36	33	12					81
(d) =	12	84	192	175	208	81			740
(b) × (c)	14	12	24	140	252	126	28		582
	36				805	402	296	163	1666
TOTALS		132	249	327	1265	609	324	163	3069

Scribner dec. C × 10 = Scribner bd. ft.

3069 × 10 = 30,690 bd. ft. = 30.7 MBF

and averages may be more accurate for trees that are not round.)

The tree-measure scale accompanying this chapter is for trees of Girard Form Class (FC) 76. This is a reasonable average FC, usually acceptable to buyers and sellers of timber on small parcels. FCs range from 65 to 90. The greater the taper, the lower the FC. The figures are derived by dividing the diameter, breast high (d.b.h.) into the inside bark diameter of a tree at the top of its first 16-ft log. Thus, a tree that measures 16 in. d.b.h. and 12 in. inside the bark 17 ft above the ground (allowing for a 1-ft-high stump) will have a FC = 12/16 = 0.75, or simply 75. Tables have been developed for each full figure of the range of FCs.

While tables provide scale data for every whole inch, even-inch tallies (12, 14, 16, etc.) generally suffice for small landowner sales.

Tree heights are measured with hypsometers and levels, both available from supply houses. For timber sales from small tracts, whole log tallies usually suffice, though tables are available for half-log gradations in heights. Only merchantable heights (not total heights) are used. The use of even-inch d.b.h. and whole log tallies is acceptable because more precise measurement is too costly in time to be worthwhile.

Merchantable sawlogs are usually those above 9 in. d.b.h. and with at least one 16-ft log to a height where the tree has a 7-in. inside bark diameter. Pulpwood merchantability usually begins at 6 in. d.b.h., and the trees must have at least two 4-ft sticks to a height of 4½ in. inside bark diameter. The popular Chip-N-Saw mills enable some lumbermen to utilize small sawlog-size trees which, for other manufacturers, are less than acceptable for lumber.

The price of stumpage—what one pays for trees standing in the woods—plays a role in the merchantability of a stand. If wood is abundant and inexpensive, buyers may consider 10-in. or even 12-in. d.b.h. trees pulpwood and may choose not to harvest for pulp any trees below 8 in. In ordinary times and for most of America, buyers must be able to depend on an *operable cut* of at least 1,000 bd ft or 4 cords of pulpwood on any acre of the tract to be harvested. Operable cuts, like merchantability, vary by region, availability of wood and the general economy.

Tree workers make deductions for obvious defects in the trees they measure and tally. Thus a 16-in. tree with a certain amount of crook or exposed rot may be tallied as a 12-in. stem. Deductions for hidden defects (rots, mainly) are made when the total volume is obtained. For conifers, 10% is a useful rule of thumb; for broadleaf hardwoods, 15%.

Trees are marked with paint as they are tallied. Guns are available for the purpose. Special tree-marking paint must be used: other paints contain ingredients that are absorbed through the bark and injure trees. Marked trees often may await harvest for several years; in that time, toxic chemicals in paint can kill them.

Many mills buy wood by weight. Conversion factors vary, depending on species, locale and company. Dead timber weighs considerably less—as much as 20%—than green wood. One set of conversions from the Doyle rule for sawlogs to green weight is

pine: 1,000 bd ft = 16,000 lb
hardwood: 1,000 bd ft = 18,000 lb

For pulpwood, the green weight conversion is

pine: 1 standard cord = 5,250 lb
hardwood: 1 standard cord = 5,600 lb

As rules of thumb, 1,000 bd ft equals 9 tons and a cord equals 2.6 tons.■

Table No. 7

Scribner Decimal C Rule—Form Class 76

Tree Diameter (in.)	Number of usable 16-foot logs 1	2	3	4
10	3	4	5	—
12	4	7	9	—
14	6	10	14	15
16	9	15	20	23
18	12	20	26	31
20	15	25	34	40
22	18	32	43	51
24	22	38	52	62
26	27	46	64	77
28	31	55	75	90
30	36	64	88	105
32	41	73	102	123
34	47	84	117	141
36	53	96	134	163
38	60	108	151	184
40	66	120	169	206

Example: A 20-in. d.b.h. tree with 35 merchantable feet (2 logs) = 25 # or 250 bd ft = 0.25 MBF.

Rules of thumb for pulpwood 6-9 in. d.b.h., four 4-ft sticks per tree, eighty 4-ft sticks per cord

Example: 100 6- to 9-in. trees = 400 sticks, 400 sticks = 8 cords

C H A P T E R 2 8

PRICING TIMBER

The high price of lumber in 1973 was not just a matter of inflation. While the price of groceries had risen almost 13% in the previous twelve months, and the price of all non-farm commodities had increased by 5½% in that period, decking-grade pine plywood jumped from about $6 to $9 for a 4x8-ft. sheet. That 67% increase in price was about the same as the rise for lumber and other processed forest products. Not since has there been such rapid escalation.

Observers of this dramatic increase are not likely to suggest an eventual return to "normal." Rather, normal will be ever-increasing prices for wood products. That prediction holds true through this writing, even though the price of stumpage dropped in the early 1980s to less than half of its all-time, mid-1970s' high.

The situation in the U.S. can be attributed to several events. The export of wood from the Pacific Northwest to Japan receives a lot of press coverage. More wood crossing the Pacific means less of it flowing to American woodyards. Besides these exports, Northwest wood competes with southern-grown pine in southern U.S. markets. This helps cut costs to the consumer of wood utilized in the warmer section of the country. More southern wood went north into the vacuum created at the retailer's yards as Douglas-fir and ponderosa pine from western forests went to Asia.

Some sales to Japan are under long-term contract and help greatly to stabilize the balance of payments for the U.S. This is particularly so for harvests in Alaska. The high cost of labor there makes even the use of high-quality Sitka spruce prohibitive in the lower 48 states. In fact, when a resident of Alaska needs to patch the step on his porch, the board he buys locally will have been shipped through the Inside Passage from a Washington or Oregon mill. It is a cheaper plank than would be one cut from the log of a locally grown tree. (Wages for Alaskan labor, not the price of the stumpage, account for this.)

This is not all bad for Americans. Many Sitka spruce logs return from Japan to North America as Yamaha pianos and high-quality guitars. Other woods sent to the Orient are shipped back in the cores of lauan mahogany doors, by far the least expensive door on the American market. Home-buyers appreciate those Philippine mahogany doors.

Logs from the Pacific Northwest are also exported to the People's Republic of China at prices competitive with those paid for timbers from the South Pacific area. This roundwood, as unmanufactured material is called, provides employment for U.S. Pacific Coast mill- and woods-workers, although considerable doubt exists as to whether the industries which own the land from which the trees are cut re-

Crown classes for black spruce in Michigan, as in this stand, help determine the value of a tree, *U.S. Forest Service photo.*

ceive appropriate compensation for the stumpage. Keeping operations humming (in the mid-1980s) at a cost in stand and stock inventories may be adequate compensation for the owners. Northwest wood moves so reasonably from one side of the Pacific Rim to the other that roundwood harvested from the southern pineries cannot begin to compete in this export market.

The federal government restricts exports of trees from national forests and private lands to control the situation. More recently, the Japanese have volunteered to limit purchases. That is probably not altruism, for vast stands of marketable Monterey pine in Chile, Australia and New Zealand are for sale at reasonable prices. Much of that timber could supplant U.S. logs now sent to Japanese mills. (Monterey pine is native to California.)

There is another side to the timber-pricing dilemma. Canadian wood moves into the U.S. in quantities and at prices which undercut sales for domestic wood products. Both Pacific Northwest and southern mills and loggers suffer economically. As most Canadian forests are owned by the federal and provincial governments, stumpage sales are controlled by governments. To maintain a high level of employment, Canadian trees are sold for much less than their value would be if they had been growing south of the 49th parallel.

Consequently, Central and South American countries purchase newsprint from Canadian mills instead of U.S. plants, and New York construction firms use Douglas-fir from British Columbia's Coast Range rather than that produced from trees growing in Washington and Oregon. In the mid-1980s, the typical cost at the neighborhood lumberyard for Canadian wood was about 15% less than for U.S. supplies. To compensate, the U.S. Administration in 1986 threatened to impose a 15% tariff. Canada responded with a 15% export tax on its softwood lumber.

U.S. trade agencies continue to remedy the situation, as U.S. residents view the matter, with tariff restrictions. Opponents of tariffs fear retaliation. Canadian imports *do* hold down forest-product prices while putting U.S. citizens out of work and keeping low the price of a landowner's stumpage.

The Jones Act of 1916 also affects wood prices. That legislation says that all shipments from one U.S. port to another U.S. port must go by an American boat. U.S.-registered ships are manned by members of seamen's unions. But wood moves from Canadian ports to U.S. ports on boats flying the flags of Third World countries. Those sailors earn much less in wages than do union members. Cheaper freight means cheaper wood at the lumberyards. It also means a greater drain on Canadian forests. That translates for the present into lower stumpage prices for U.S. timber.

As "Dust Bowl" storms by a strange trick of fate compounded the 1930s Depression's toll upon the nation, so the severe weather of 1972 and 1973 accompanied timber price increases in those years. Floods and heavy rains kept loggers out of the woods for weeks at a time. Mills ran on short shifts, log inventories at mill decks were measured in hours, not months, and specialty plants—like box factories—shut down. When logs are in short supply, wooden boards are priced high.

Quick and costly action in that period was also required to save flooded lumber from fungi-caused stain and decay. Cleaning and kiln-drying is the best way; chemical treatment without kiln-drying is of little benefit. These techniques are costly.

There is another reason for ever-increasing lumber, plywood, and paper prices. Compliance with federal and state regulations for pollution control and occupational safety costs money. Environmental studies must be prepared, reviewed and approved before ground can be broken for wood-processing plants. Machinery to reduce air and water contamination has to be purchased and installed. Equipment to reduce noise from power saws in the woods also must be provided for the worker.

The effects of regulatory measures will continue to increase the cost of wood, both on the stump and at the mill. DDT, for instance, is among the cheapest and most effective insecticides. Federal foresters once claimed it was needed, as an emergency measure, to control the tussock moth in an outbreak of from 300,000 to 400,000 acres of conifer timber. The Environmental Protection Agency withheld its permission, preferring to depend on a virus which kills the insect. It usually provides control on a three-year cycle. If the EPA had not been overruled because of a provision in the law, more expensive salvage logging would have been necessary.

The federal government's actions add to the construction industry's unpredictable cycles. Washington permits or restricts low-cost housing and federal building, adding to the tension. Industrialists pay for that tension, charging it finally to the user of wood.

Foresters often complain of the "agrarian mentality" that says that forests are useful only to hold together the land until it is needed for food crops. Others emphasize development, financed with adequate capital, promoting forestry only until the land is needed for potential homesites, airports, and factories.

These two "higher uses" of woodlands may necessitate land conversions in the same area at about the same time; cropland developed for housing means the more-fertile forests become farms. Factories breed homes as airports breed factories. It must follow, as the cart follows the horse, that when land for the production of any goods or services becomes scarce, the value of that which is produced on that land increases.

In many states, the *ad valorem* tax payments to the county for schools and roads continue to rise. In one heavily-forested southern state, property valuation more than doubled in the seven-year period ending in 1970. Unfortunately, where taxes are higher, commitments to forest production are lower. Only so much money at any one time is available for landowners to put into growing timber. The larger the bite for taxes and associated expenditures, the lower the balance for insect and disease control, girdling cull trees and protecting lands from fire. The consumer here, as most everywhere else, pays.

Aggregating properties, an innovation encouraged by one state forestry service, is one approach proposed for use on a national scale in order to help assure the owner of timber an appropriate profit for his efforts. By this method, lands held by many owners would be leased to an organization qualified to manage them. A group of farmers might employ a professional forestry consultant, a conservation association, a school of forestry or a cooperative established for the purpose.

Owners would be paid "ground rent" from government treasuries until the first harvest, perhaps in about 15 years. By then, the managed stands should be productive. However, the lessee, or management team, would be required to follow up over a long period, perhaps 40 or 50 years.

Ground rent of $10/acre/year is a suggested assessment. The only outlay of the owner would be one-half of the *ad valorem* taxes. All other costs would be borne by the management team which would, in turn, reimburse the landowner 10% of the value of the timber harvested after costs are covered for the initial rehabilitation of the woodlands.

While higher stumpage prices are passed along to consumers, they also provide funds which landowners can use for management practices. As stumpage prices fluctuate for sawlogs and plywood in most of the commercially forested parts of the country, landowners remain puzzled. How much money *will* be available for improving the woodlands?

The timber glut that unexpectedly appeared about 1980 in America cannot last long, for the current shortage of high-quality stems for sawtimber and plywood have made straight, clear boles a prized commodity. If $150/1,000 bd ft for trees that take 40 years to grow is discouraging to the taxpaying landowner, perhaps $400 will not be.

With a good market, the landowner would be encouraged to intensively manage his forests to provide for others from his store of resources. That is what farming the small forest is all about. That, too, is what conservation—the wise use of resources—is all about. ■

C H A P T E R 2 9

TAX BREAKS AND TAXING TIMBER

"Will the nation's forests be adequate?" That is the question asked by resource economists, government officials, and industrialists. It is a question that should be asked by users of paper and by buyers of homes. The inquiry is also appropriate for the owner of small tracts of forestland. Increasing costs in the future of items made from trees could reflect a potential scarcity of the supply of wood.

Concern for the security of the nation's woodbasket has prompted Congress to provide some tax breaks for the small timberland owner. These incentives are intended to encourage landowners to invest in the improvement of the growing stock in their woodlots.

Forests are considered a national resource—necessary for security as well as for a comfortable living. Because the time it takes for trees to ripen for harvest is long, federal intervention is considered appropriate, even by the most conservative capitalists. Congress thus has provided a favorable investment climate to stimulate owners of small tracts to restock their lands, thin stands of trees, eliminate the culls and protect the woods from insects, disease and fire.

Estates, certain corporations and partnerships are also eligible for the deduction.

To do this, the nation's legislators, at the end of 1980, wrote into law (PL 96-451) a deduction for amortization of the cost for reforestation. (Think of amortization as the gradual death—*amort*—of a debt.) Certain restrictions are placed upon this provision. The deduction for amortization only can be for a maximum of $10,000 spent by a taxpayer in any one year. It is a deduction, not a credit. The expenses may be spread out over no more than seven years (eight tax years—the last six months of the first year and the first six months of the eighth year). They must have been made in 1980 or later, the benefit being retroactive to that time. The law is generous in allowing the deduction for care of tracts as little as one acre, provided the land is for commercial timber production.

Planting costs allowed to be deducted, in addition to site preparation and seedlings, are for seeds (for direct-seeding operations), labor, tools and equipment depreciation. If the property is sold within 10 years, these amortized deductions are paid back to the federal treasury through "recapture" as ordinary income. (Money reimbursed to landowners for forestry incentives practices [FIP] under federal cost-sharing programs handled by the the Agricultural Stabilization and Conservation Service

Earlier, lands where trembling aspen "quakies" grew were taxed at low rates. Discovery of the tree's usefulness for paper-making in the Lake States has raised *ad valorem* assessments, *U.S. Forest Service photo.*

[ASCS] is generally included in the taxpayer's gross income.)

To illustrate, a 40-acre abandoned agricultural tract owned by Mr. Smith has lain idle for years and grown up in brush. It is a good site for pines. The State Forest Service prescribe-burned the area on January 10, 1981, to prepare for planting in February of the same year. Here is the tally of costs:

Site preparation	$ 160
Seedlings (40,000) @$25/M (thousand)	1,000
Labor for planting @$50/M	2,000
Tools (dibbles, pails, etc.)	300
Supplies (moss, fuel, etc.)	250
Depreciation on truck (prorated)	100
Total	$3,810

When April of the next year rolls around, the Smiths will have a deduction of $272.14 or [(3810/7)/2]. The total cost is divided by seven because the law stipulates this is the number of years over which the amortization may occur; the division by two in the first and eighth years occurs because the law says the period begins on July 1, even though the costs may have been incurred in January. The same would be the case if the expenses were in December.

In the second through seventh years, the deduction will be $544.29 or (3810/7). The eighth year it will again be $272.14. At the end of the period, the Smiths will have deducted their entire costs for site preparation and planting. How much they get back from IRS, however, depends on their tax bracket.

Owner Smith could treat another tract in 1982 and another in 1983. The IRS deduction, however, will allow for no more than $10,000 in costs for any one year. Mr. Smith cannot spend $11,000 on a 200-acre tract this year and claim a deduction for the extra $1,000 next year.

There is another item in this tax legislation that the small landowner should know about. It is called the investment tax credit. This is a 10% credit to a maximum of $10,000 in any year, even for expenditures covered under the amortization deduction. It is subtracted from taxes owed. There is a prime exception. Equipment already subject to the depreciation tax break cannot qualify for the credit. Both the amortization and the investment tax credit were retained in the 1986 tax reform act passed by Congress.

Good records are necessary to support deductions which can reduce taxes. With those records, the timber-selling landowner can show costs, which are subtracted from gross income to arrive at a net taxable income.

Costs of tree-farming generally fall into three categories: capital expenditures, operating expenses and expenses of the timber sale.

The land survey, boundary-marking, corner monuments, maps and the initial cost of the timber are capital expenditures. The land on which the trees grow is a capital asset, the purchase cost being a capital expenditure and therefore not entered on the timber account. Preparation of planting sites, with bulldozing and weed-tree injection, and all costs involved in tree planting or direct-seeding, are also capital investments. Such costs are capitalized, and appropriate amounts are deducted from income when timber is harvested to determine the taxes owed. The owner should keep a separate accounting system for each tract.

Operating expenses cover the recurring costs of maintaining and protecting a forest property on which timber is grown. Expenses may include salaries and wages paid (but not to the owners, even if they do the work), equipment repair and maintenance and the price of expendable, inexpensive tools, such as axes, saws, compasses and logging chains. Operating expenses include interest, taxes on the timber (again, not on the land on which it grows) and taxes on heavy equipment. Depreciation of big equipment, such as tree-planting machines and tractors, is not an operating expense. Separate accounting is required.

Most operating expenses are charged as current costs and deducted from income. On properties where there is no current income or where deductions are not itemized on Internal Revenue Service Form 1040, certain operating expenses—including taxes, interest and insurance—may be capitalized as carrying charges.

The third type of deductible cost relates specifically to a particular sale just completed. Consultation by foresters, timber cruising, tree marking, scaling and advertising the stumpage are typical expenses of a sale. These, too, are deductible from current income.

There is another deduction. It is for casualties, of which the forests of America have more than enough. Ice, windstorm and fire are examples. To the insurance company, these are Acts of God; to the IRS they "result from an identifiable event of a sudden, unexpected or unusual nature."

To the sorrow of landowners, attacks by insects and disease carriers are not considered casualties for tax purposes. This is so even if the insects and fungi invade as a result of weakened tree conditions due to an earlier fire or storm.

A further word about casualties for tax-computing. Many landowners lose out on casualty claims because they cannot prove the claim. For timber growers, this means frequent inventories so that when the "identifiable event" occurs, a subsequent cruise of the timber remaining enables the owner to verify the loss. IRS requires that the expenditures to date be calculated and that the volume of timber prior to the loss be determined. From these data, the cost per 1,000 bd ft or per cord is determined. This is another good reason for frequently making inventory cruises, as the "unexpected" event may wipe out the forest, leaving little trace of its former value.

One of the few places where inflation is not legal is in the evaluation of casualty losses. A landowner who has invested $5,000 in a stand of pines cannot claim more than $5,000 if the grove is wiped out by lightning-caused fire. This occurs even though that stand may have doubled in value due to growth and inflation. Or another example: claim for total loss of a pulpwood-size forest grown from seedlings at an investment cost of $5,000 could not exceed this figure, even though it was worth $20,000 just before the disaster.

Landowners receive payments from the federal government for rural environmental conservation practices (RECP). They may also receive incentive payments for improving forest management for fiber production (FIP). Both kinds of payments are considered ordinary income, reported like wages, in the tax year the checks are received. Landowners' matching costs will, of course, be tallied as a capital investment. They get their reimbursement for this when the timber is sold. A landowner receiving FIP payments may be able to claim both the government share (75%) as well as his cost (25%) as a capital expenditure.

Landowners pay taxes to local entities by two measures: *ad valorem* and severance. *Ad valorem* refers to current value, usually of land and timber. As land increases in value due to demand and as trees grow with time, this tax levied by local and state entities increases. Sometimes *ad valorem* tax concerns only the land, and not the timber on it.

Severance taxes are paid to local and state jurisdictions as the trees are harvested. This tax may be by value or by volume of timber cut. The next few years will likely see *ad valorem* taxes diminish and severance taxes increase to compensate for the losses of the former. Governments prefer the continuity and dependability of the former; landowners prefer severance taxes, not paying the fee all of the years they wait for trees to grow to merchantable size and highest value. Combinations—small *ad valorem* taxes paid annually and severance taxes paid at harvest time—provide a happy compromise.

On that land another stand will grow to provide wood and services for another generation. It is to provide for those future Americans that Congress encourages high timber yields through incentive payments and tax breaks for woodland owners. (The capital gains tax, intended to provide relief for individuals who, perhaps only once or twice in a lifetime, sell property or other capital assets at a profit which would otherwise be subject to high regular-income tax rates, was eliminated in the law which took effect on January 1, 1987. As I write, Congress is being encouraged to amend the statute to again enable this break.)

Computation of Taxes on Sale of Timber From a Farm Woodlot
(MBM = 1,000 bd ft)

Timber Sold		
100 MBM (thousand board measure, or feet) @ $30 per M		$3,000
Sales Expense		
Forestry consultant	$100	
Timber cruising	75	
Marking trees and scaling logs	150	
Total sales expense	$325	
Depletion Allowance Basis for Gain		
Amount originally paid for timber (exclusive of land value)	$1,200	
Surveying cost and legal fees (chargeable to timber)	128	
Forestry consultant expense (incurred when property was purchased)	100	
Site preparation and tree planting	200	
Capitalized taxes and operating expenses*	217	
Total capital invested	$1,845	
Amount of timber when land was purchased		110 MBM
Timber growth since land was purchased		80 MBM
Total timber before current cut		190 MBM
Unit depletion (1845/190 = $9.71 per MBM)		
Basis for gain on sale (100 M × $9.71 per MBM = $971)		
Less sales expense plus basis for gain ($325 + $971 = $1,296)		$ 1,296

*Applicable mainly for taxpayers who do not itemize deductions; for others it is advantageous to claim these expenses in the year they occur.

The more significant tax laws affecting forestry, as of January 1, 1987, are these:

The 1980 "Packwood Bill," providing for reforestation tax incentives;

The 1981 Economic Recovery Tax Act, providing reduced tax rates;

The 1982 Tax Equity and Fiscal Responsibility Act of 1981;

The 1984 Tax Reform Act, causing no changes in timber taxation; and

The 1986 Tax Reform Act, eliminating capital gains rates, but retaining reforestation incentives and expensing (with "passive" limitations).

Note: The writer is obliged to state the usual disclaimer regarding taxes: Landowners should consult tax advisors and lawyers for counsel on their particular situations.

CHAPTER 30

APPRAISING FOREST LAND

Land is almost like money; it changes hands so fast. Yet sellers, and less often buyers, are not aware of the real value of the real estate they offer for sale. Real property is not called real by accident; it is among the most reliable indicators of value.

Unfortunately, statistics are hard to find for forest land values over the years, in contrast to farm real estate. For the latter, which may be a gauge of woodland worth, values have been increasing since 1933, after falling by some 60% in the early Great Depression years. In times of depression, land is a better deal than other kinds of property, such as stocks and bonds; it may also be a better investment in good times.

One-hundred-year-old stands of valuable wild black cherry, like this woodland of high-quality trees in Pennsylvania's Allegheny Mountains, encourage high land appraisal values, *U.S. Forest Service photo.*

Others have written alarmingly of the effect excessive *ad valorem* taxes could have on the value of forested lands. In some areas, this tax may exceed half of the value of new wood laid down each year. It is a cost that must be borne annually, not when the harvests are made. Therefore, this tax often encourages sudden sales of timbered land.

It was the taxes on cutover land that, in part, caused its abandonment following the harvest of the virgin forests—from 1860 until 1940—throughout much of the eastern and southern parts of the country. It was this tax that led to land purchases by the U.S. government from willing sellers in the 1930s, giving rise to new national forests in the East, the South, and the Lake States.

Some states provide loan money for veterans who want to buy land. However, the maximum amount available for each veteran is limited. Thus, the 25 or 30 acres of potential forest land that one could purchase with these low-interest loans would be inadequate for a commercial farm woodlot in most regions of the country. In one state, the Veterans' Land Board permits a group of five loan applicants to each procure a tract of at least 10 acres in a single block. While this is generally regarded as a means for developing a residential subdivision, the board could allow the purchase of the land for timber-producing purposes. Any parcel less than 100 acres in one undivided tract would be difficult to manage to attain maximum value for the timber growing on the land.

Forestland appraisals begin with an accurate description of the tract, preferably the whole courthouse record, giving the acreage and showing land lines, corners, roads, structures and rights-of-way for powerlines and pipelines. Useful, too, but often missing from these plats, are streams, springs and ponds. Shoreline distance, and the extent of ownership out into a lake or up the slope at a lakeside to a waterline contour, should also be shown. Shape of the tract could be significant: extremely odd shapes impede management. Aerial photographs, obtained for virtually all of the U.S. through county Agricultural Stabilization and Conservation Service (ASCS) offices for a reasonable fee, can be used with mapped overlays to show timber types and physical features.

Additional kinds of information needed for appraisals are distances to oiled or paved roads and whether pavements are farm-to-market or higher class. Timberland values depend on mileage to sawmills, plywood plants, pulpwood rail yards, wood-treating plants, pole mills and professional forestry assistance. The latter is important because consultants charge mileage, and the client is probably paying for travel time. Industry and government foresters whose responsibilities include providing free assistance to woodland owners are more willing to help if the land is fairly accessible. Unlike residential property, the distances to and the quality of schools, churches and hospitals do not enter into appraisals of timberland.

When all of this is on paper, the appraiser goes to the woods. After leaving the paved road, the condition of the forest road is noted. Is it all-weather, graded and graveled, or will loggers and forest managers expect to spend time digging out of mud holes or waiting for better weather?

Is the tract visibly marked? Signs are inexpensive and should be nailed to fence posts (not to trees, please), giving the owner's or manager's name and phone number. Opportunities for good timber and land sales are often missed because potential buyers are unable to readily locate owners.

Does *that* fence go with the property? Is it on sturdy pressure-treated creosote posts? Fence posts of post oak, baldcypress, and cedars are not worthwhile when one considers the high price of labor, wire, and staples for erecting, maintaining, and replacing a fence. Pressure-treated creosote posts last up to 40 years.

Fences in poor condition detract from land values; they should be removed. Tripping over barbed wire is a psychological stumble for a potential buyer of land or timber. Appraisers and consulting foresters charge owners for torn trousers and boots. Well-blazed and painted boundaries and easily spotted permanent corners, with correctly marked witness trees, are an asset. Searchers for corners may readily find an ankle-high or buried pitchy coniferous wooden "stob," as pins and stakes are often called, that had been driven into the ground 30 years earlier because the witness trees properly pointed to its position. One may find these stobs of lightwood (the core of the heartwood of virgin pine stumps, boles, or branches) buried by earlier surveyors under piles of rock that have protected them for decades from fire.

Having walked the land lines, the appraising party—seller, buyer, real estate agent, or professional forester hired to do the job—will likely next look at the stands of timber. The appraiser will need to know the species, number of trees per acre, volumes in those trees, average diameters, the range of diameters, and the number of 16-ft logs or 5-ft pulpwood sticks in the trees. For some deals, a simple cruise giving this information will do. Landowners must avoid encouraging a best-guess estimate by appraisers. Both will regret a less-than-cursory examination.

Values determined by appraisals will depend upon quality of the timber. Appraisers must know the going rate for logs of various grades and be able to estimate for the tract the average and the extremes in quality. Stumpage values may vary by 1,000% per 1,000 bd ft, depending upon tree size, number of branches, straightness or crookedness, form (whether boles are cylindrical or conical), forkedness and defects caused by diseases and insects.

The soil must be examined. A sharpshooter spade is the best tool for quickly digging holes large enough to view, although augers and sampling tubes serve the purpose. In the absence of trees, it is the soil the appraiser depends on for estimates of the potential of the land to grow trees of various species.

Site index (SI) gives the average total height of the dominant and codominant trees when 50 years old. For pines, SI may vary from 50 to 125-plus. A general figure is about 80. With curves like the ones shown in Chapter 26, this measure of land productivity can be given for trees at any age. Neither the number of trees to the acre, their rate of diameter growth, nor their girth affect site index. It is influenced principally by surface and subsoil texture, drainage (both internal and external) and soil nutrients.

Where trees have been cut, killed by fire, toppled by storm or destroyed by insects or disease, evaluation is aided by soil-site tables. An abbreviated ex-

Table No. 8.
Site Index* for Loblolly Pine in Southeast Texas

Subsoil Texture	Surface drainage	Depth to Subsoil (inches) 0 to 10	10 to 20	over 20
		Site Index**		
Sands to	good	L	M	M
loamy sands	poor	M	M	H
Clay loams to	good	M	H	H
light loams	poor	H	H	H
Silty clays to	good	L	L	M
heavy clays	poor	M	M	H

* S.I. = the average total height of the dominant and codominant trees on the land at age 50.

** L = Low (site index of 70-80), M = Medium (site index of 80-90), H = High (site index of 90-100).

ample of one for loblolly pine in the West Gulf South is shown. Foresters have available similar tables for many species and regions.

The appraiser should be able to distinguish soil texture classes (sand, silt, and clay) and to identify mottling (colors caused by oxidation, hydration and reduction of iron coatings on the soil due to drainage conditions). A forestland appraiser will know that some trees grow well in sites where agricultural crops do not. But the landowners should not depend solely on that. A rule of thumb for forestry is that the best sites are mesic; neither too wet nor too dry; moist, but well-drained.

From the site and stand information, the productive potential of the land can be ascertained. Will it grow 50 cu ft or 85 cu ft/acre/year? Will that be worth $6 for pulpwood or $150 for sawtimber?

Past use of the land will aid in estimating value. For instance, if it is in the South and was once in row crops, the site would be an unlikely one for planting slash pine. Annosus root rot destroys many trees of this species used to reforest old-fields. If the land is subject to flooding, certain broadleaf species would die if inundation exceeds two days. If it was previously clearcut, what species were on the land prior to the harvest? Would these be appropriate for reforestation? Helpful, too, would be knowledge of the original virgin forest type and whether wildfire (look out for the mischievous neighbor!) can be expected to occur.

Recent appraisal values of other forest lands nearby and accounts of actual sale transactions accompany the report. Determining this information, called the market-data approach to appraising, may be time-consuming and, therefore, costly.

One must distinguish between appraisal values and actual sales. Some land will sell for more than its appraisal, perhaps to block-in a property or for maintaining "the family place." Often, too, land goes for less than the appraisal, for sellers who need money promptly may be anxious to be relieved of a property. Appraisal values should be rounded to the nearest $100 for the whole tract or $5/acre.■

CHAPTER 31

EVALUATING TIMBER

Increases in timberland values of 500% over 10 to 20 years were common throughout the country in the 1960s and 1970s. Some of the inflationary trend occurred because people preferred timberland to other kinds of property ownership. Indeed, many with excess financial resources use land purchases as a hedge against inflation and a place to tuck away capital. In so doing, they contribute to further inflationary spiraling of timber values.

Such rapidly rising land values do not make sense unless the crop produced on the land more than pays for the inflationary increase in value. An aspiring timberman cannot buy land today, compute the purchase costs annually in compound interest and come out ahead. For the cash a man may use for buying forest land at $800/acre would, if invested otherwise at 7%, reward the investor with $4,342 in 25 years.

To figure your own situation, use the formula

$Vn = Vo(1.0p)^n$, where

Vn = value of the land at the end of the period,

Vo = value of the land at the beginning of the period,

p = the interest, and

n = the number of years (or the period) the investment will run.

Bottomland hardwood timber and land values are judged by bark characteristics; the Nuttall oak on the left depicts a stem of poor vigor; the slightly smaller tree on the right shows, by its bark, high vigor, *U.S. Forest Service photos.*

For the previous example, Vn = \$800 times 1.07 to the 25th power, or \$800 times 1.07 multiplied by itself 25 times. Bankers, agricultural county agents and foresters have tables that do the calculating.

The rich farmland bought for \$200 an acre in 1932 theoretically was worth \$2,080 in 1987 if an average annual compound interest rate of 5% were used for the calculation. Cutover timberland that sold throughout the nation for \$5/acre in 1935 would, 50 years later (again theoretically), provide the seller with \$52, apart from inflation, at 5% interest compounded annually.

Yet the timber land's 1987 value, with a good stand of timber that originated following the mid-thirties' cutting, and well-cared for since that time, often exceeds \$1,000. Half of this value would be the worth of the wood standing on the land as trees.

Fortunately for fiber (and food) consumers in this country, landowners seem never to have carried inherited land values "on the books." Nothing we read or live in would be so cheap if we, as consumers, were charged for the investment in the land on which the raw material for paper, lumber and furniture is produced.

Having established the current value of the land, we now consider the value of the timber growing on the land. Important measures for determining the value of standing timber are basal area and volume. Tree farmers need to know about them.

Basal area is a reliable estimate of stand density. Measured in square feet, it is the sum of the cross-sections 4½ ft above the ground (d.b.h.) of all of the trees on an acre. To get the idea, visualize every tree cut, leaving table-high stumps, and then add together the areas of all of those round table-tops and you have the basal area. An example: a tree 12 in. d.b.h. has a basal area of 113 sq in. or 0.79 sq ft. The calculation follows:

$$A = \pi r^2$$

where $\pi = 3.1417$, and

$r =$ radius of the bole at b.h.

$$\begin{aligned} \text{Thus } A &= 3.1417 \times 6^2 \\ &= 3.1417 \times 36 \\ &= 113/144 \text{ sq ft} \\ &= 0.78 \text{ sq ft} \end{aligned}$$

Well-stocked stands of southern pines have 100 to 120 sq ft of basal area. Poorly stocked stands may have 60 or less. Fewer large trees have the basal area of many small trees. Because many tree sizes are represented in a typical woods, this single value conveniently tells a lot about a tract of forestland.

For wood volume determinations, measurements are made in board feet, cords and tons (see Chapter 27). Accurate cruises are essential; the greater the percentage of the land tallied, the greater the accuracy. Small landownership cruises run from 10% to 100%. The greater the percentage, greater also is the cost of the cruise.

Tree and stand age, rate of growth and evenness of growth are also important for evaluating timber. Ages of trees in a stand may vary, as for eastern and western hemlocks, or, as for the southern pines, all the stems may be the same age. Some trees may be young, mixed in with a stand that is principally composed of old-growth.

Size is often confused with age: trees of the same age and in the same stand growing close to one another may vary from 2 to 20 in. in diameter. The larger ones somehow had an early advantage that blocked some sunlight to the lesser ones. Bigger trees, too, compete for soil moisture and nutrients to the detriment of smaller ones. Failure to realize age and size leads to diameter-limit cutting, the bane of forest management.

In many states, standing timber is real property; trees become personal property when felled. Trees "on the stump" are in a category with land as surface rights and not with minerals as subsurface rights. While in some states the law may not so require, buyers and sellers of timber may wish to deal with agents licensed by state real estate commissions when having timber, as well as timberland, appraised or exchanged. On the other hand, many states have few licensed real estate brokers who have adequate knowledge of either timber or timberland values to handle such matters: simply having a license is not adequate for handling these transactions.

A notable problem with timber values is in the swings of the pendulum. Within a couple of years, stumpage may double in price and then return to its earlier level. And uncertainty is what American businessmen—whether forest landowner or corporate executive—cannot tolerate. Spending by landowners for timber stand improvement is drastically reduced in times of uncertainty.

The landowner may know with certainty that the value of the trees on his land rises abruptly and dramatically when they grow from pulpwood into sawlog-size trees. For much of the country until the mid-1980s timber recession, growth from 8.9 in. d.b.h. to 9.1 in. increased greatly the value of such trees. Nine inches is the dividing line between these product classes. With depressed markets, however, 11 and 13 in. often mark the division between pulpwood and sawlog products. While small sawlogs will not bring premium prices, their worth will far exceed that of the best pulpwood markets.

As a rule of thumb, 1,000 bd ft contains a little more than two cords of wood. (There is no precise conversion factor due to the variation in size and shape of roundwood pieces. Some companies arbitrarily rate 4 cords as 1,000 bd ft.) Thus, small-pine sawlog stumpage, priced at $60 to $70/1,000 bd ft, is of considerably more value than two cords of wood at $6 to $10 a cord. The landowner may find it preferable financially to delay a thinning for a year if, in that year, significant numbers of stems will grow into the 10-in. sawlog class (9.0-10.9 in.). Another helpful rule of thumb to aid in making this decision is that for each $1/cord increase in price due to inflation, the increase in value for sawlogs will be $10/1,000 bd ft (MBF).

Pulpwood values did not increase appreciably from 1956 until 1971. Indeed the upswing in 1972 was no greater than the down-swing in the mid-Sixties. The price for this class of stumpage remained rather steady through the mid-Eighties. For sawtimber, the low point was reached in 1963 and again in 1983. Since then, the price rise—when the wrinkles in the curve are ironed out within each calendar year—has been consistently upward for trees destined for sawmills and plywood plants. Continuation of this trend will enhance the value of the land on which that timber grows. Local fluctuation is not uncommon.

Prices paid within regions, and sometimes for a state, are published at frequent intervals by independent agencies. County agricultural agents, banks and consulting foresters know where to obtain that information.■

CHAPTER 32

COSTS AND INCOME

The imagined cost of doing business keeps a lot of owners of small tracts of forest land from managing their timber. Forest landowners are categorized by federal statisticians as "small" if they own less than 1,000 acres of woods. Yet, it is these people who are expected to produce much of the future fiber needed in the U.S., according to the prophets responsible for keeping tabs on the nation's forest resources. As ownerships change, more professional people with analytical interest become the overseers of the woodland resource.

This chapter itemizes the outgo and returns of a hypothetical case. It is fiction only in that it is not the data from a single woodlot. The information could have been obtained from any farm forest in which forestry is practiced. The pros and cons of these practices may help landowners decide what they must do to most wisely manage and utilize the resources in their care.

Our imaginary tract is 100 acres of pine-hardwood forest in the East. The average acre has 4,000 bd ft (4 MBF) in pine and 1 MBF in a mixture of oaks, hickories, and other broadleaf trees. This volume of wood is, on the average, in 25 sawlog-size pines (above 9 in. d.b.h.) and 10 hardwoods (larger than 13 in. d.b.h.) rather unevenly spaced on each acre. Average spacing between trees is about 35 ft. Basal area of sawlog-size trees amounts to 110 sq ft per acre.

The pines in this stand came in following clearcutting and fire 40 years ago. Though the pines in the forest are of a single age (even-aged), stems vary from 3 in. in diameter at breast height (d.b.h.) to 20 in. Smaller stems are suppressed, suffering from the competition of larger trees for light, water and nutrients. The larger trees, for some reason, got off to a faster start, remaining indefinitely the dominant stems in the woods.

After 25 years, the tract was thinned, taking out diseased stems, insect-infested trees, some suppressed ones and leaners. That harvest, mostly for pulpwood, removed about 8 cords/acre in some 100 pines. The hardwoods entered the understory following the establishment of the conifers. None has

A mixture of Christmas trees and timber trees—as plantations and natural forest—on a Tree Farm in Maryland provides income for its owners, *American Forest Council photo by Lester A. DeCoster.*

been removed in thinning or otherwise harvested.

Some of the land is upland and dry; some is bottomland and of high site quality. All has been burned by wildfires, but without any significant damage. Rather, the fire helped hold back weed hardwoods and bramble growth.

A good map is essential, showing boundaries, logging trails and streams. The timber, too, should be shown as pine, hardwood, or pine-hardwood stands. Some classification of age (if it should vary) or tree size is helpful. Basal area notations help: the relationship of this statistic to growth, and therefore earnings, is strong. For some counties, a Soil Conservation Service (SCS) soil map is available. (A book published by the SCS for each of the counties of the country for which a modern soil survey has been made is one of the best bargains available. The volume describes the soils, notes their productive potential for various crops—including trees—and contains fold-out aerial photographic maps. No landowner should be without this tool.)

These maps can then be referenced to an aerial photo obtained from the local Agricultural Conservation and Stabilization Service (ASCS) office. If available at the same scale, an overlay of the map on the photo is useful. Preparation of the map, assuming land lines have been run and corners established, could take about two man-days for our hypothetical tract.

A more precise timber cruise will document the data presented above about stand volumes, basal areas and tree sizes. For 100 acres, this could be a 100% cruise in which the d.b.h. of every tree is measured. The cost of such a cruise is estimated at eight man-days. The price would be reduced significantly if a partial cruise is made. The trees on as little as 10% (considered adequate for tracts of this size for planning purposes) of the land could be measured in 50 1/5-acre plots randomly spaced over the land. Some foresters prefer to run strip cruises, sampling say 10 acres, in lines evenly spaced on the tract. The cost: 4 man-days.

One might assume that a 40+-year-old second-growth stand is ready for its second thinning. On this occasion, about 100,000 bd ft, mostly pine, is marked for removal. The going rate for this sale of 80,000 bd ft of pine (80 MBF) is $100/MBF or $8,000. The 20 MBF of hardwoods, valued at $25/MBF, bring another $500.

Sale of timber for thinning does not bring near the price that clearcutting does. Unfortunately, it is often impossible to find loggers willing to move equipment into tracts for partial cuttings. Consequently, clearcutting may be an economic necessity, though not necessarily the best conservation practice. Under some circumstances, clearcutting may be the best practice, both ecologically and economically.

Should thinning be considered inappropriate, a stand of this size, age and vigor justifiably could be designated for a final cutting. As selection and shelterwood regeneration harvest systems may not work for pines in this woods, either the seed-tree or clearcutting system must be used (see Chapters 5 and 6).

With the seed-tree method, five to 10 pines/acre are selected to provide seeds for the new forest. These 500 to 1,000 stems on the 100-acre tract, among the best in the stand, will be harvested later. Some will be lost to lightning during this period. The price paid for stumpage will be less than maximum because of the relatively few trees for which the logger will need to move in men and machines for cutting, skidding, loading and hauling.

Maximum income is obtained with clearcutting, the stumpage rate for a stand like the one described being about $150/MBF in the 1980s. Income for this small landowner forest could thus gross $60,000.

All logging costs are borne by the buyer. Seller's expenses are for the timber cruise, map, contract writing and sale services.

Maximum values will be obtained by advertising for sealed bids. Even small landowners should follow this procedure. The additional 60 to 90 days required to consummate the sale ought to be well repaid in a better deal. That big companies pay the best prices is a misconception. Bids should go to all possible buyers.

The signed contract should, among many things, limit the time in which the timber can be removed. In this stand, growth could exceed 10%/year, allowing extra income for the buyer as a reward for delay. Inflation, too, encourages purchasers to contract promptly and cut later, perhaps a year or more, when the value of the wood is anticipated to be higher. The trees may add significant increment in that period. Some stems, on the other hand, may be lost to lightning strikes, insects, disease and wind.

Woods are sometimes left a shambles, creating criticism for clearcutting as a proper silvicultural system, and wasting the site. Regeneration must be considered when the sale is planned, for every year an acre of forest land is out of production can be counted as $10 to $20 lost to the owner, depending upon the price for stumpage when the tract will again be cut in the distant future. And at appropriate interest rates for the rotation (the time interval until final harvest), that can be costly.

Site preparation and planting, averaging $125/acre, may be required to establish the new forest. Pushing over unmerchantable material, windrowing it with tractors, and burning the debris could cost $70 or more. Machine planting of 600 seedlings at a spacing of 8x10 ft may cost $50 for the equipment and operator and $30 for the pine planting stock. Hand-planting would cost more than $30/acre. Industries may spend well over $100/acre to chop the brush and logging slash with large crawler tractors. Shearing and piling prior to planting cost $75/acre.

Other management costs include the use of brush-control chemicals. Mist-blowing herbicides to release seedlings may amount to $25/acre. Release of selected individual stems with tree-injectors could be a little less. Neither of these treatments would be required on the whole tract. Perhaps 15 acres might be captured by vines and shrubs. Perhaps, too, at age 10, 20 acres may require weed-tree control of encroaching low-grade hardwoods that overtop the pines.

Other costs include insurance, boundary-line maintenance and interest on loans for forest practices. Taxes take a major bite; they vary from $1 to $20/acre/year for forest land. A landowner may spend as much on property tax (averaging $4/acre/year in the South) as on all silvicultural practices for a moderately intensively managed woodland. Owners of tracts of this size often employ forestry consultants. This adds another item to the outgo column. It should also enhance the income column; for wise management of natural resources, what "conservation" is all about, should more than pay its way.■

CHAPTER 33

TREE FAILURES COST MONEY

The haste with which the injured (or their estates) now rush to the courts for redress is readily observed by reading the news from even the smallest communities. Too often the injuries have been caused by falling trees or limbs. As judges and juries appear to be sympathetic to the injured, rural landowners may benefit from an understanding of their responsibilities before becoming a litigant in a civil trial at the county seat. Even to be an innocent defendant is costly in time and attorney expenses.

Courts will generally side with defendants who can show "good faith." They will be more sympathetic to the landowner who provides abundant evidence of his being aware of the danger of "risk" trees, periodically examines his premises for these hazards, and removes those that could threaten bodily or property injury. On the other hand, a landowner who has shown by his acts that he is aware of these dangers but does nothing to correct the situation probably is worse off than the one who pleads, and can prove, his ignorance.

The weight of a pinc 14 in. in diameter is about 2,000 lb. A tree that size is likely to have its center of gravity 30 ft above the ground. Perhaps a half-ton of wood hangs perilously overhead. Hardwoods are heavier; Douglas-fir and other soft softwoods are lighter. The fall of such a weight may devastate that which was beneath it.

What should landowners look for? Risk trees include those that lean more than 1 ft horizontally for each 6 ft of height. They topple with the wind, the center of gravity being to one side. Leaning stems may be root-sprung, the ball of soil in which the roots are bunched breaking away from the earth and severing the stabilizing roots in the process. Trees may be root-sprung that are not leaning. To identify this condition, the landowner looks for a crack in the ground circling the trunk several feet from the tree base.

Trees with considerable rot in the interior—weakening the stems so that they are easily broken somewhere above the ground by wind—are usually readily noted. They may have wounds at the base of the tree caused by mechanical injury or fire. Well above the ground may be openings in which squirrels nest, red cockaded woodpeckers make their home, or in which bees build a hive.

A spindle-shaped or round canker on the boles of pines indicates the presence of a *Cronartium*-caused disease. Cell growth of the living cambial layer is stimulated by the fungus' hormone production, but the wood is greatly weakened. Breakage occurs at this point. Branch cankers, covered with orange spores in the spring of the year, may weaken the main stems by the growth of mycelia—minute hair-like strands that grow up and down the wood within the bole. Where the canker occurs on a branch within a few inches of the trunk, infection has probably spread widely within the tree. The spindle is readily identified by its fusiform-shaped deformity,

Tree failures that injure people and property often cost landowners money, *Mike Legg photo.*

the round canker is crinkled, appearing like an animal's brain. Thus they have the names C. (for *Cronartium*) *fusiforme* and C. *cerebrum*.

Loggers call spike-topped trees widow-makers. The dead top of the terminal branch protruding through a green crown is a spike top. Stag-headed trees are also dangerous, their many bare branches—appearing like the antlers of giant elk—easily broken by the wind.

Other problem trees are those that have suffered ice damage. The whole top may be gone, exposing a wound to rot-causing fungi and further deterioration. Lightning strikes and exposed roots are also signals of impending risk. Water-saturated soil, surface-eroded soil, and tight compact soils, in which stabilizing roots cannot penetrate, are other clues to watch for.

Forked trees tend to break more frequently than single stems. Gnarled, aesthetically pleasing, old scraggly boles may most readily respond to the pressure of wind.

Many trees in the forest grow in loamy or sandy surface soils. Underlying the coarse-textured zone is frequently a hardpan or claypan, a layer of dense soil which even tough tap roots cannot pierce. Under this condition, the tree may appear to be stable and resistant to high winds. As it grows, however, its center of gravity rises further from the ground and upward on the heavy bole. Thus, the chance of its being uprooted increases. Blowdown is especially serious when hardpans are within 2 ft of the surface of the soil.

Dead branches are a hazard wherever people walk or congregate. Removing them may be costly, even more expensive than felling the tree. The fall of a branch from the top of a tree is not a sign to cut out the living top. Many a boulevard has been ruined by this overkill. Removing the green top has only served to provide exposed tissue for the ubiquitous spores of decay fungi.

Courtroom testimony may use cross-sections of fallen trees to prove or disprove good faith by a landowner. The case of a tree that fell across a road during a heavy rain is an example. Did the tree lean, and, if so, should the owner have been responsible for its removal?

Lawyers, juries and judges will want to know whether a wind-toppled tree leaned before it fell. That can be determined with some degree of accuracy by the presence or absence of compression wood or tension wood in the bole. The former occurs in coniferous softwoods and the latter in the broadleaf hardwoods. A leaning pine produces compression wood on its lower side, as it does on the bottom side of branches. The rings will be closer together, and the wood denser and darker, than on the upper side.

Tension wood forms in deciduous trees on the upper side of the leaning bole or branches. Determination may require microscopic examination of the wood cells. Tension wood is notorious for splitting.

Both of these kinds of "reaction wood" are formed in the tree in an effort to restore the stem to an upright position. The tree literally reacts to reduce the stress placed upon it. The occurrence of both kinds of reaction wood can be observed in cores taken with increment borers.

Because a tree is old and large and has withstood many winters, storms and fires does not mean it is destined to persevere for years to come. Just as risk is increased with an increasing number of trees and with increasing tree age, it is also influenced by increasing tree size. The probability of failure is gradual to a point, say 20 in. in diameter for water oak, and then rises abruptly with increasing diameter growth.

All trees, of course, will eventually die, and the risk of failure attributed to unhealthiness increases with time. Thus, there is a subjective evaluation in assessing risk. Individuals, agencies and companies will likely look for an acceptable level of hazard rather than the absolute absence of hazard. So, too, will a jury. To have no hazard is to have no trees along the roads or trails of a forested tract or the pathways leading to it. Reducing the risk by removing potentially hazardous trees is like increasing one's auto insurance as further protection from liability.

One state highway agency requires, as the result of injury litigation, that all of its roads be examined twice yearly in a search for risk trees. The concern exists even though the risk trees are on private property adjacent to publicly owned rights-of-way. A court required a small community in the early 1980s to pay a high price for an injury to a child on a municipal playground because of a falling limb. Even a logger won an appreciable settlement, just as the trial was recommencing following a recess, because of a tree top that fell on him while on the job in the woods. That potential litigants may be less likely to sue landowners of small forests is wishful thinking in these days of busy courts and active lawyers.■

CHAPTER 34

TIMBER SALE CONTRACTS

In a market economy, the seller tries to get the best deal he can. In this system, the buyer has the right to bargain for his best interest. This bargaining makes the system work.

In the sale of timber, bargaining is concluded when the contract is signed. Unscrupulous activity is curtailed, and honest people can know where each stands on points that could be contested later. Not only do timber sale contracts minimize lawsuits, they also preclude ill feelings. Equally important, contracts should encourage good conservation, giving assurance that land earmarked for timber production will remain for that purpose. A professional forestry consultant familiar with contracts is recommended for small landowner sales. However, in the event none is available, the following ideas should be helpful.

The landowner must know what he has on the land. Timber volume is determined according to a particular scale, such as the Doyle or Scribner, from the d.b.h and number of merchantable logs in each tree. Knowing the number of merchantable trees per acre and the average tree diameter for each species represented, the average volume per tree is determined. Other information the owner might record is the maximum and minimum merchantable diameters of the trees to be cut, the basal area per acre and the vigor of the trees.

The landowner then must determine what he wants left after the harvest. The cut could be a thinning, leaving perhaps two-thirds of the stand; a seed-tree harvest, leaving 10 high-quality trees per acre for regenerating the forest; or a clearcut, after which nothing of monetary value is left.

What is now on the land and what is to be retained following the logging operation become part of the contract. Minimum dimensions of each tree sold are noted, including the d.b.h., length of log and the proportion of the wood that is free of rot.

The care of residual stems may mean long-term profit or loss. Careless loggers scar remaining trees while skidding logs. They run down seedlings and saplings with tractors and trucks, and hang cut trees into those not designated for sale. Careless loggers will also puddle soil, harvest trees when the ground is too wet and foul creeks with debris and mud.

To help avoid this, the contract should call for a bonus to the logger—not necessarily the buyer—for minimal damage to the stand and the soil. The contract should note that the bonus—perhaps as much as 5% of the contract price—would be determined at the discretion of the landowner or the landowner's forester. The bonus could be for any portion of a stipulated maximum amount. Lawyers say that a third party cannot be bound in a two-party contract. Consequently, buyers do not bother to tell their loggers "what's in it for them" if they do a careful job, as the bonus does not benefit the buyer. Thus, the landowner must assume the responsibility to know when logging will begin and be at the site to explain

Cindy and Marshall Stacy produce pulpwood, sawlogs, and Christmas trees. Their Tree Farm won for them the industry-sponsored Outstanding Tree Farmer award, *American Forest Council photo by Lester A. DeCoster.*

the bonus to the logger. Hopefully, loggers shown a copy of the contract will take the seller's (or his agent's) word.

Contracts should spell out responsibility for damage to the site. The documents ought to note the buyer's liability for the actions of his contractors. Unnecessary erosion, fence destruction, road washout, broken culverts, wildfire, high stumps and cutting of unmarked trees are examples. The latter are usually paid for at a rate of double the sale price of the stumpage.

It is currently appropriate, and someday may become law, for buyers to restore cutover lands to productive condition. Thus, the contract could require felling of unmerchantable material on clearcut tracts and windrowing and burning of logging slash and other debris. Broadcast burning or chipping of logging debris also may be included in contracts.

Controlling brush with chemical applications, planting seedlings or direct-seeding treated seeds would likely be handled by a separate contract with people equipped and qualified to do this work.

If individually selected trees are to be marked, the means (by paint) and where (at breast height and ground line) should be stated. Diameter-limit cuts should be discouraged, but if one insists on this method, the limits should be stated precisely. For instance, "trees 12.0 in. in diameter and over at 10.0 in. above average ground line, the average ground line determined by measurement on the north and south sides of trees."

For tracts that have been partially harvested, and then seemingly forgotten, it may be impossible to resell the trees that were not cut by the logger. Thus, a performance bond may be required of the buyer. Banks handle these documents. The bond should be part of the contract at signing time. Payment for all or part of the timber, depending on the volume, is made known to both parties by this paper. If partial payments are made, the contract should note the volume of timber in each sale unit. A map showing these designated units becomes a part of the document.

Such a map should outline logging roads if needed, landings (where truck loading is done), and, if available, timber cruise information. The basic map could be drawn from field notes recorded at the courthouse. Land grants and other location data are shown.

Time limits are important. Timber sold at a current-year price and left to grow two or more years may have added 10% or 15% to its value. Only for bad weather and other "acts of God," as noted in the contract, should delay be permitted. Some buyers anticipate storing wood on the stump where it continues to grow at a faster rate than the interest on the bond at the bank. An adequate bond also helps to check this habit.

Subcontractors may not be accountable, since they did not sign a contract with the seller. If they are to be permitted to log the land, the document should so indicate, specifically itemizing the contractor's responsibility, as noted earlier, for those he hires or otherwise designates as his agents.

If the sale is to be by log-scale, rather than tree-volume, the location of scaling should be noted. It may be at the landing in the woods (where logs are loaded on trucks) or at the mill. The particular scale stick (again perhaps Scribner or Doyle) is stated. Wood volume may also be recorded by weight, the scales usually located at the mill. Certified scalers should send duplicate copies of measurement tickets to the seller if payment is not made in advance for the entire stumpage volume or for a designated portion of it. These pay-as-cut sales should be avoided whenever possible.

In this day of environmental concern, Best Management Practices (BMPs), as determined by state and federal governments, may become a part of the contract. These relate to non-point source pollution (such as sheet erosion) and point-source pollution (such as gullies forming at log-landing sites). As governments specify what these practices are to be, loggers, foresters, landowners and contractors will be expected to comply. Landowners might anticipate governmental decrees by stating these requirements in sale contracts (see Chapter 24).

BMPs will raise the cost of wood to the consumer. (They may also lessen the price the seller gets.) As an example, in one western state every stream, both intermittent and permanent, must be walked following logging. Any debris in that stream bed caused by the harvest must be removed in order for the creek to continue to run clean and unimpeded. That single BMP adds $10 to the cost of each 1,000 bd ft of lumber for the home builder and home buyer.

Following are a suggested invitation to bid letter and a timber sale contract. As the author is not licensed to practice law, users of this book should consult local attorneys for appropriate documents.■

Letter proposing sale of timber:

J.J. Greenwood
Consulting Forester
address

Date
To the person addressed

Dear Sir:

Acting as agent for ______________________________, this is to solicit sealed bids for pine and hardwood timber on about 10 acres of land belonging to this party near ________________ in ________________________ County, ____________.

Our estimate of the trees to be cut, in board feet, Scribner scale, is approximately 90.2 MBF after deducting 10% for hidden defect from the gross volume. Stand and stock tables are attached.

While the tally was made with care, potential bidders should ascertain volumes for themselves. The volume above is in 644 trees, averaging 159 board feet (gross) per tree. External defects were considered in making the cruise.

The number of trees and the volume do not include 23 otherwise merchantable trees, marked and flagged with a yellow painted S, which are reserved for seed.

A standard contract is to be used with these exceptions:

(a) double stumpage to be assessed for damage to the seed trees, at the discretion of the forester retained by the owner,

(b) Fences are to be restored, and

(c) $300 deposit, to be returned when (a) and (b) above are satisfied.

Bids will be received until 5:00 p.m. on October 15, 198___, at the address above, noting on the envelope "WB Timber Sale."

The successful bidder will have one (1) year in which to remove the timber.

Bids will be on a lump-sum basis, payable upon signing the timber deed.

A map attached shows the location of the land on the east side of US 01, 1.3 miles south of the Farm Road intersection. The stand may be visited at your convenience.

Owner reserves the right to reject any and all bids.

Please call me at (___) ___________ for more information.

Sincerely,

____

Consulting Forester

STATE OF ____________________
County of ____________________

TIMBER DEED

For and in consideration of the sum of __________________________, cash in hand paid, and in other good and valuable considerations, the receipt and sufficiency of all of which are hereby acknowledged, ______________________________, hereinafter called "SELLER" do hereby convey and warrant unto ______________________________, hereinafter called "PURCHASER," all the pine timber as hereinafter described in tabular form and estimated to be ________ thousand board measure (MBM), net volume, Scribner, on the following described land:

In the __________________________ Survey in ______________________ County, ______________________, about ________ acres bounded on the east side by a graded road, on the north side by a fence, on the southwest side by a stream locally known as Crooked Creek, and on the west side by a recently surveyed line near a barbed-wire fence. Note: The fence that approximates the west boundary line should be ignored. See map attached.

The land is located on the west side of U.S. Highway 101, 1.3 miles south of the intersection of that highway with State Road 94.

All timber sold under this agreement has been marked with yellow paint spots on the bole of the tree and below stump height.

For any hardwood (broadleaf) trees containing merchantable timber which is cut by the Purchaser, his employees, contractors, or employees of contractors, except for removal of the pine timber, Purchaser shall pay Seller double the price of stumpage currently paid for the class of material said trees contain. [*Or* Purchaser may cut and use such small hardwood timber as may be necessary for bridging, roadbuilding and logging.]

No unnecessary damage shall be done to the ______________ designated and otherwise merchantable seed trees, so marked, or to the soil. Double stumpage is to be assessed for damage to the seed trees, at the discretion of Seller's retained forester.

Seller shall pay direct to logger(s) ____________________ dollars ($________) or a portion thereof for logging the land in such a manner as to do minimal damage to the soil or to the residual seed trees or hardwood trees. The allocation of this award shall be solely at the discretion of the Seller's professional forester. [N.B. Lawyers may argue about this paragraph. See earlier.]

Purchaser shall have the right of ingress and egress on, across, and over the land owned by Seller for the purpose of logging the timber conveyed herein. Fences required to be cut for access shall be permanently repaired when such access is no longer needed.

Purchaser shall gain access at the designated (yellow flagged) place on the graded road, then follow the flagging for the location of a log landing (deck), also shown on the attached map. Purchaser may use any part of the tract for skidding and bunching.

Unless extension of time is granted in writing by Seller, the timber sold under this agreement shall be cut and removed from the above-described land by November ______, 19___; title to any timber sold under this agreement and remaining on the land described above after such deadline or any extension thereof shall revert to the seller.

A $300.00 deposit will be returned when fences are restored and seed-tree damage, if any, assessed and deducted.

Purchaser agrees and warrants that it will at all times indemnify and save (hold) harmless Seller against any and all claims, demands, actions, or causes of action, for injury or death of any person or persons, or damage to the property of any third person or persons, which may be due in any manner to the operations of Purchaser upon this land.

IN WITNESS THEREOF, THE SIGNATURES of the parties to this agreement on this, the ________ day of November 19___, are affixed.

____
Landowner Jones, SELLER

Witnessed:

____
Buyer Company, Inc., PURCHASER

CHAPTER 35

PROFESSIONAL FORESTRY ASSISTANCE

Professional foresters deal with the sciences basic to the growth of trees and the business of managing those trees for the benefit of people. The practice of forestry sometimes embraces the art of conservation for the use of natural resources apart from fiber-producing trees that are found in wild lands. These additional renewable resources include water, herbaceous plants eaten by grazing animals, wildlife and its forested habitats and esthetic landscapes enjoyed by humanity for recreation.

Forest management practices need not be limited to wild lands, those tracts never subjected to the plow. Foresters care for individual trees in urban settings and for the stands of trees that comprise municipal parks. They also protect from wildfire, insects and diseases those forests which have been set aside for non-commercial purposes in national and state parks, in wilderness areas and in wildlife refuges. Establishment and care of many of the nation's "Third Forests"—plantations of conifers and high-quality hardwoods on lands once cultivated or cutover—are the work of their hands. These Third Forests, planted on lands from which virgin stands and, later, mature second-growth stands have been harvested, are found on both private and public lands.

Professional forestry assistance enabled the owner of this naval stores stand in the Southeast to properly plant and manage the slash pine plantation for fiber and for gum.

Some dozen states license or register professional foresters. These state-approved resource managers—like physicians, attorneys and other professionals—subscribe to a code of ethics and hold memberships in organizations of their peers.

For foresters, the ethics code has been formulated by the Society of American Foresters (SAF). This national organization—founded by Gifford Pinchot, "the father of forestry in America," and his colleagues in 1900—provides services for citizens through chartered state or regional groups. These charters require members to subscribe to the parent organization's code of ethics.

As a tax-exempt and non-profit organization, neither the SAF nor its state subsidiaries are permitted to lobby or to accommodate any self-serving interests of their members. Rather, the professional association is responsible for furthering the cause of forestry in America by creating opportunities for free interchange of views, disseminating knowledge to the public as well as to its members and "fostering a spirit of comradeship among foresters."

To enhance the management of the nation's forestlands, the SAF proposes policy statements for review and adoption by its members and council. Members of Congress, legislative committees and others use these position papers for developing appropriate pronouncements and commentary on matters involving natural resource use and abuse.

Recent position statements involve growth and harvest relationships for a state's commercial forests, management practices for the national forests, continuing educational programs, the need for prompt regeneration of cutover land, incentives for encouraging landowners of small tracts to improve management practices, reclamation of surface-mined sites, publication of soil survey information for forested counties, maintenance of trees in urban settings, encouraging voluntary use of guidelines for forest practices by all landowners and appropriate use of prescribed burning.

Policies considered by SAF members deal with appropriate ownerships for forestland, resource inventories, land use, forest protection and management of soil, water, fish, wildlife, rangelands and recreation areas. Issues also addressed by SAF touch on mineral extraction, research, education, taxation, financial investments for forestry and international opportunities and responsibilities.

Professional foresters who are SAF members have at least a bachelor's degree from an accredited forestry school. In addition to basic instruction in the sciences and in business, their education covers such diverse fields as entomology, tree pathology, soil science and hydrology. Some specialize in managing for wildlife; others for recreation, for the production of clear, clean water, or for enhancing range lands. In order to be technically well-versed stewards of natural resources, many foresters participate in continuing education programs sponsored by SAF, universities, government agencies, industry or privately funded associations.

Foresters serve the public as consultants to landowners and industry, as educators, and as public servants responsible to government agencies and, therefore, to the citizenry. Many foresters manage the nation's vast inventory of industrially owned commercial forest lands. Some buy wood for domestic and foreign corporations. Others are employed by citizen and industry issue-oriented associations.

Consulting foresters manage land, appraise and

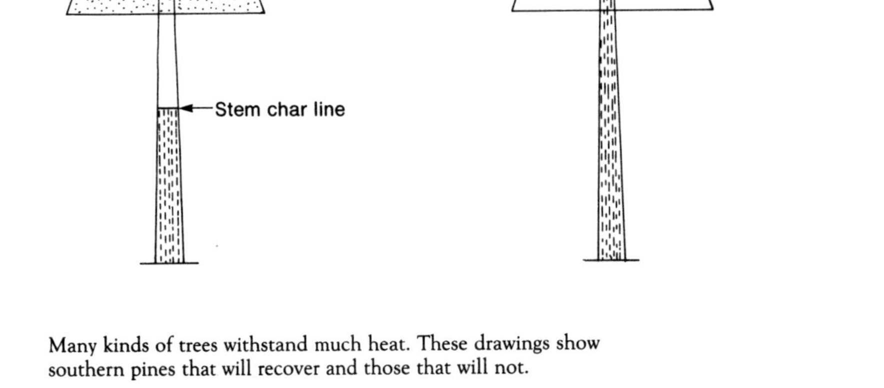

Many kinds of trees withstand much heat. These drawings show southern pines that will recover and those that will not.

sell timber and solve technical problems. Educators instruct college and high school students and landowners (the latter through a network of state and federal bureaus and industry-supported programs).

Government foresters are employed by many agencies. Among them are the federal departments of Agriculture, Interior, and Defense; state forestry and parks organizations; counties; and cities. Their work involves the whole realm of environmental resource conservation, including aid to owners of small properties.

For both public and private woodland owners, professional foresters are expected to obtain maximum benefits in goods and services from the land for which they are responsible. Owners' goals determine this objective within the limits of a conservation ethic.

Conservation, originally a forester's term and defined as "the wise use of resources," involves an understanding of ecology. This term has been in the forester's glossary since the turn of the century. As those who "study the interrelationships of living things to each other and to their environment," foresters became the original environmentalists. Their work, involving the whole of environmental and ecological relationships, also pegs them as resource economists who recognize that management must provide materials and services ". . . for the greatest good of the greatest number in the long run."

More than 500 forestry consulting firms are providing services for industry, government, and landowners. Some firms employ more than 10 foresters, while others are single-person operations. Forestry may be part of a larger business where engineers, clerks and statisticians are also on the staff. Many consultants are people retired from other tasks. Some are part-time self-employed specialists.

Timber management, land surveying, recreational planning and real estate appraisal are significant functions of the consultant. Writing environmental impact statements and management plans for the small forestland property is often an important part of their job.

State forestry services, the Soil Conservation Service and agricultural cooperative extension services of land-grant universities also provide assistance to individual landowners. The American Forest Council, through its *American Tree Farmer* magazine, and the Forest Farmer Association, through its *Forest Farmer* magazine and annually published *Manual*, continually update information for nonindustrial managers.

Other organizations with helpful publications include the American Forestry Association, the Soil and Water Conservation Society, the Western Forestry and Conservation Association, state forestry associations, and the Society of American Foresters.■

C H A P T E R 3 6

MANAGEMENT PLANS

Every forested tract should have a management plan. Those lands for which federal and other funds have been allocated must have such a plan. Basically, a plan indicates that the owner has given thought as to what should be done in the forest, when it should be done, by whom it should be done and how it should be done.

These plans are not "set in concrete." Weather, economy shifts, social changes, tax reappraisals and legislation may alter the purposes for which an owner retains a property. In that event, the plan for the land may require alteration.

Plans need not be complex. One for a 100-acre forest may be but two or three typewritten pages plus a detailed map. The map should show compass bearings and lengths of all boundary lines (noting adjacent ownerships), as well as interior structures, trails, springs, streams, utility rights-of-way, fences, distances to principal roads and acreages of compartments within the tract. An aerial photograph may accompany the map. A large-scale blow-up of one of a stereoscopic pair of photos is a useful field tool. A simple, inexpensive pocket stereoscope enables observers to visualize topography and tree heights.

On the map and the photo also may be recorded timber types, insect-devastated spots and areas in which the owner schedules (by year) planting, thinning, or harvesting. Contour maps, available for most of the U.S. from the Coast and Geodetic Sur-

Aerial photographs, obtained at reasonable cost from U.S.D.A. Agricultural Stabilization and Conservation Service offices—in almost every country seat—facilitate management of a woodland.

vey, also show the lay of the land. In spite of persistent admonitions, some mappers continue to fail to include a north arrow (specified as to *true* or *magnetic* north), a distance scale, the date of the plat's making, and the name or initials of the map maker. If boundary lines on the ground do not coincide with fences, this should be noted on the map.

Landowners find a forester's compass a worthwhile possession. It differs from most compasses because east shows to the left of north. Hence, the user looks along a line of sight, after the dial is moved to the desired direction, so that the direction one wishes to go appears at the tip of the north needle. To go east, the dial is moved so that east coincides with the north-pointing needle, sending the user east. With a little experience, one gets the hang of the instrument's usefulness.

Management plans should include stand and stock tables. These, respectively, note the number of trees and the volume, by species, in each diameter class. Management plans include stand basal area, tree ages, vigor of trees, site indexes and soil description. The working plan notes the volume to be removed in each harvest, the number of trees, the basal area and the year for each harvest. Rotation ages and cutting cycles are noted.

An owner's purpose in holding land is a foremost consideration for the design of the plan. Some people hold forestland solely for income production from timber sales. Others want to preserve a natural area. In between are those who wish to have a weekend retreat, perhaps to have some income while retaining the pristine beauty of the tract. Another needs hunting land; someone else aspires to a financially rewarding investment as land values increase. The desire for a hedge against inflation should not be overlooked in longterm goals and objectives.

As management progresses, records should show volumes harvested and financial income and outgo.

In summary, a management plan outline, modeled after one in an old text (*The Management of Farm Woodlands*, by G.H. Guise, McGraw-Hill Book Company, 1939), remains appropriate.

Management Plan Outline

I. Objective of owner
II. Regulatory policies of governments and owner
III. Map(s)
    A. Forest
    B. Compartment areas
    C. Boundaries and neighbor ownerships
    D. Interior detail
IV. Basic mensurational data
    A. Stand tables
    B. Stock tables
    C. Areas of stands
    D. Vigor of stands
V. Management procedures
    A. Cutting cycles of stands and/or compartments
    B. Rotations of stands and/or compartments
    C. Silvicultural treatments
    D. Cutting plans
    E. Planting plans
        1. Species
        2. Locations
        3. Stock description and availability
IV. Records
    A. Volumes harvested
    B. Planting completed
    C. Expenditures
    D. Receipts
    E. Financial Objectives
    F. Other

USES OTHER THAN FOR WOOD

The forest as a multiple-use resource

CHAPTER 37

NATURAL FOREST FOR ESTHETIC APPRECIATION

"In wildness is the preservation of the world." So wrote Henry David Thoreau in the mid-1800s. Though as simplistic as saying "sugar is sweet," it was a high-sounding statement, sure to trigger the enthusiasm of anti-exploiters who would in time become the leaders of the conservation movement in America. The obviousness of the wildness-equals-preservation relationship gave rise to few debates. A century of conservation leadership, only now concluding, ignored it; for, to these later people, conservation is the wise use of resources.

An old-growth northern hardwood forest. In the autumn, as the foliage turns to dozens of hues of color, each species exhibits its own peculiar shades, changing from day to day and throughout the day, *U.S. Forest Service photo*).

Now, again, Thoreau's line is recited. However, currently it is not to encourage conservation but, rather, preservation. The preserves so established are designated wilderness; and wilderness locks up resources from human uses—save two.

One is recreation, and specifically limited recreation. Eventually only some confined hiking, camping, and canoeing is permitted; for unlimited recreational use of these lands results in unpreserved preserves.

The other use of preserved sites is for observation, including amateur "bird-watching" and professional research. Future studies of areas now set aside will enable people to understand nature's way of doing things and, therefore, how best to manage the forest.

Those opposed to wilderness in the East and South argue that this land will be needed for growing wood. Housing starts of about 2 million units a year in the U.S. make a significant demand upon renewable forest resources. Even when starts slipped by 10% to 15%, as in the late 1970s and early 1980s, due to higher mortgage-interest rates, the harvest of sawtimber exceeded volume growth in some locales.

Then, in 1977, pressured by preservationists and a call from the White House, the Forest Service initiated a second Roadless Area Review and Evaluation (RARE II). A new study was underway.

In the 18 months allotted, Forest Service officers nominated all tracts of appreciable size that had few roads through them and invited the public's reaction to setting these lands aside as part of the National Wilderness System. Units in the purchased national forests in the East and South—in contrast to the national forests in the West which were taken from the public domain—would be an extension of the "Wilderness East" Act of 1975. That legislation led to RARE I.

Under that law, the federal wilderness could be sites that had been cutover or farmed before purchase in the 1930s. No longer need "wilderness" be lands that man had never used, or in natural unimpaired condition. Thus, under consideration for wilderness designation were forests currently managed for fiber production. Within the national forests, of course, are sites of unusual character deserving wilderness classification.

The contest by letter-writing, as required by law, was tabulated by a centralized computer. Nationally, 75,000 favored no further wilderness designation, while 14,000 sided with the preservationists.

Foresters often consider wilderness a management dilemma. How to satisfy opposing publics is a question that escapes an answer. A ready solution may be removal of small properties from tax rolls and transfer to reserve status.

Some states have short hiking trails, scenic routes that range in length from half a mile to many miles over all kinds of terrain. They may be located on lands held by either individuals or corporations. These woodland trails may be hiked all year, though rainy weather may make access to a few difficult.

The trails serve many uses. One may provide a short respite for a weary traveler as a courtesy of the sponsor. Another is a conscious effort to win support for the manner in which the owner cares for the land, believing that support of one's stewardship of the forest can best be obtained by observing "on the ground" management practices.

Trails are trod by all kinds of folks. Scouts earn patches for hiking them. Families trek the footpaths for afternoon outings, often picnicking among the managed trees.

Some paths are marked with signs that tell about nature. For instance, one tells of a loblolly pine infested with red heartrot, where the red-cockaded woodpecker makes its home. High in the tree, the visitor is told, the brownish conk, or fruiting body, of the fungus which causes the rot is seen protruding like a shelf from the bole. Or by a swamp, its waters coated with green algae, grow magnolias with their large white, sweet blossoms that made Dixie famous. Rocky streams, too, occur along some trails decorated in spring by flowering dogwood blooms.

Again, the sojourner may be informed about the pesky pine beetle, an insect having little concern for the care of the trails. Listening closely, one may hear the wood-consuming insects chew away in perfect cadence. Next the dead tree becomes part of the soil; nature's recycler is never stilled for long.

Hikers may be told about fire which has always played a role in the ecologic changes occurring in the woods. The blaze is evidenced by char on tree trunks, charcoal particles in the soil or the occurrence and absence of certain kinds of plants. Swift-footed white-tailed deer abound in some areas, depending upon the food supply, and turkeys may have been introduced.

A southern live oak woodland: the leaves remain green throughout the year and the clinging Spanish moss drapes over its branches.

Or dig a hole. Here one sees the organic matter contributed by decayed plant and animal life incorporated into the mineral soil. Above this zone may be a thin layer of pure organic matter, not mixed with soil. It may be a thin surface mantle where a warm, moist climate encourages rapid rotting of dead material. Below the zone of incorporation is the surface soil free of organic matter. This may be from 6 in. to several feet deep. At its lower limit, one may see an abrupt transition from a coarser-textured to a finer-textured soil. Hence, a sandy loam may be underlain by a sandy clay, or a silty loam may rest above a silty clay.

Color, too, tells a story. Well-drained lands are red; those continuously water-logged will be dark, sometimes blue. Mottling—the mixture of many shades of reds, yellow, and blue—occurs where soil water fluctuates. These colors form because of chemical reaction of the hydrogen and oxygen in the water and in the soil's air pores with the iron that coats every soil particle. The red is oxidized iron, like the rust of a nail; the blue is reduced, the opposite of oxidation, as in a swamp. Yellow is often attributed to hydration, a sort of in-between situation.

But human history as well as natural history is found along the paths. One tells about Long King (I suppose a lanky fellow), an Indian chief who, as a cohort of pirate Jean LaFitte, at least in legend, roamed or hid nearby. Ghost towns of the 1880s, even before commercial logging really got underway in some locales, are memorialized. One lane travels by the site of an early oil well, where oil still seeps into surface "springs" in which Indians bathed and from which pioneers gathered grease for wagon wheel axles. A couple of paths follow old logging railroad beds that wind among trees now as tall as those for which the rails were laid to facilitate the harvest. Another visits the scene of an early-day river port, when clear, deep waters had not yet been silted-in with the sediment of civilization's activities.

Perhaps the site along the trail was once a cotton field, now returned to forest. If so, the soil profile will show as an abrupt line at the depth of repeated turnings of the plowshare. This plow sole may have recorded a hundred years of previously furrowed land use.

Owners of small tracts of land may consider the appropriateness of such ventures. Tax breaks may be available. Certainly such land usage provides a service to others. Some states have laws that release owners from liability for injuries in the event of accidents. Vandalism and wildfire may also be the calling cards of an unappreciative public.■

C H A P T E R 3 8

MANAGING FOR WATER

"Water, water, everywhere, and all the boards did shrink; water, water, everywhere, nor any drop to drink." Coleridge's "Rhyme of the Ancient Mariner" tells about the plague of the lack of drinking water on a ship at sea. Recall the water famine occurred because the albatross had been mercilessly slaughtered. It could be a prophesy of our time in which we, too, cry, "Water, water, everywhere, nor any drop to drink" because of the merciless destruction of watersheds through poor stewardship of the land.

Water, like bread in Biblical metaphor, symbolizes a life-sustaining essential. Our concern with water is for man's physical needs. It is a concern with but 1% of the earth's water. It is the water which is recycled and hence renewable. Like the aluminum can, it can be used over and over again. But it must be collected and channeled appropriately. This requires careful management of forested watersheds.

While most water is a renewable resource, some recycled water may require geologic time for its renewal. For all practical purposes, it is mined. That is the case for the part of the Ogallala Formation underlying much of the Great Plains and from which farmers obtain irrigation water. The Ogallala, cut off by eroded river channels, leaves forested landowners with little to do about conserving its water. (Fortunately, little of the surface above this aquifer is forested. Ogallala recharge is so slight that proposals have been made for pumping into it water collected under forests to the East.)

Cutting poorly formed trees of no commercial value, like this post oak, reduces water loss from the soil to the atmosphere through foliage transpiration. Conserved water is later released from springs and seeps to streams.

The average annual rainfall and snowfall equivalent in the U. S. is about 30 in. Seventy percent of this precipitation is utilized in evapotranspiration by plants, including trees. As forests generally grow on lands receiving more than the mean amount of precipitation, woodlands show more runoff and ground recharge than nonforested areas. About 10 in. of rainfall a year occurs as runoff or accumulates in the soil. This 10 in. needs to be carefully protected, especially because much forest is on steep, easily-eroded mountain slopes.

A few more statistics: an acre-foot of water is worth about $3,000 when used industrially. It is worth about $300 when marketed for recreation, about $30 when used in agriculture, and from $3 to $20 for the growth of trees in the forest. Hence, rain used by the forest to grow trees may be worth only about 1/1000 of what it is in the manufacture of forest products. Nevertheless, without that $3 to $20 worth of water, there would be no forest and no wood for industrial raw material.

The U.S. will see strategic shifts in forest production in the immediate future. Forestlands of this nation may be looked upon as a mammoth farm. As the farmer obligates the "south forty" for corn, the "north forty" for range, the bottomlands for cotton, and the hills for a woodlot, the Rocky Mountains a decade or two hence may be primarily water-production forests. So, too, the Southern Appalachians, leaving for the southern Coastal Plain and the high-rainfall forests of the Pacific Northwest the production of the nation's wood fiber.

In the Appalachians, foresters have measured the production of water that is relatively free of sediment as a result of harvesting timber. In the Coweeta Hydrologic Laboratory watersheds in southern North Carolina, rainfall at the summit measures about 120 in. a year. The equivalent of 17 in. of that rainfall is utilized in the transpiration of the hardwood trees on those slopes. Thus, if one cuts all of the trees, but leaves them lie on the ground so as not to disturb the soil and cause erosion, the amount of water passing through a weir at the base of the watershed is increased by the equivalent of 17 in. of rainfall annually. Those 17 added in. of rain from any number of acres—if continuously maintained without tree cover—will go down to the valley, where the rainfall averages 80 in., for textile manufacture. And then on to the Piedmont and the Coastal Plain, where 50 in. is typical, will flow water to be used for man to drink and industry to consume.

Long before multiple-use was a federally established principle for forestlands, foresters were interested in techniques for involving the land in more than a single use. They wanted to know the damage that cattle cause on steep watersheds. Grazing for domestic purposes is no longer of importance there; the cows have left the scene. The principal multiple-use problem now is how to obtain the maximum amount of clear, clean water while also growing wood fiber for the future.

To do this means silviculturally changing the natural vegetation. On the high, cold ridges in the eastern mountains where the soil is deep, post oak and low-grade pines occur naturally. On the ridges and just below them, pitch pine prevails. The intermediate mountain sides are covered with valuable oaks and maples, while in the moist coves grow yellow-poplar. Under intensive management for both fiber and water, landowners harvest the post oak and low-grade pines, replacing them on the ridges with high-quality trees such as white pine. In the coves, yellow-poplar is catered-to over the many "weed" species which enter ecologic succession on these high-quality sites. In the interval between harvest and establishment of replacement species, an increase occurs in the quantity of water discharged from the land through springs.

Management of land for water production often should take precedence over other uses. The Carrizo sands, for instance, outcropping in Texas in a crescent shape over 1½ million acres, are the source of water for two cities and a paper mill. The sands, dipping more steeply than the land surface toward the sea, also underlie another town. At that point the aquifer rests below sea level and, hence, is saline. All of the outcrop is above salt-water intrusion.

Landowners need to know precisely where formations like these occur. The Carrizo, for example, is an outcrop of extremely uniform sand that is slightly finer in the surface than in the subsoil. The vegetation on these lands becomes significant if water is an important commodity. This is shown by rules of thumb for the rates of infiltration of rainwater into surface soils. A freshly cultivated field requires 30 min. for a liter of water to disappear into the surface of the soil. Under a pine forest, it takes about 3 min., while the water disappears in a hardwood forest in 30 sec. A heavily-used recreation area—one adjacent to a picnic table or a fireplace—requires 30 hrs.

If aquifers like the Carrizo are to produce maximum amounts of clear, clean water for outflow from the underground storage reservoir, these lands should be in hardwoods. This is the case even if the species are weed trees with low or no income to the owner. Taxpayers from public treasuries will eventually subsidize owners of such lands in order for them to be encouraged to grow vegetation that will maximize infiltration and subsequent water storage.

Surface storage of water may be undesirable from a long-range standpoint. Land of the highest quality is taken out of production for food and fiber, and evaporation is excessive. Perhaps the only advantage to surface reservoirs for water supplies is the alternate role of a lake for recreation. Creation of waterfowl habitat may be another. Subsurface storage has its economic advantages in that land is not taken out of fiber and food production. Much of the nation's water is supplied from underground aquifers; the fact that 25,000 wells are dug each year in a single state suggests the significance of subterranean storage.

This chapter has been about water in the woods, telling some case histories. The management of water in a particular forest will depend upon many criteria, the combination of which will describe that forest. Its owner will need to evaluate the pros and cons for the particular set of circumstances and, then, prescribe management practices most likely to result in optimum use of available water for the growth of trees and the good of men.■

CHAPTER 39

WILDLIFE PRODUCTION— DEER

In the heat of argument, landowners and their foresters are accused of destroying wildlife in the process of producing maximum yields of high-quality wood. Some argue they cause biological deserts with monoculture plantings. Others say they fail to practice multiple use, that their interest is timber-oriented forestry.

Multiple-use was never meant to imply that every acre should simultaneously produce wood, water, forage, esthetically pleasing scenery, and wildlife. Abundant game and abundant fiber can be produced on the same land. Wood and wildlife are seldom mutually exclusive. They grow well together, enabling intensive double-use of the forest.

Landowners interested in developing favorable habitat for all kinds of wildlife should leave corridors of uncut timber that connect clearcut areas. These "stringers" sometimes are left even through clearcut tracts. Such bands of trees serve as cover for the passage of animals. High vegetation along stream courses is retained. Cover is as important as food for many kinds of wildlife.

Jagged boundaries around clearcut areas are encouraged. This provides more "edge," the transition zone in which wildlife abound. Here animals feed on food plants in the opening and, if spooked, quickly retreat to the forested canopy. Clearcuts that follow contours, rather than precisely squared blocks, provide much edge. So, too, do creeks and old logging roads when they serve as harvest boundaries. Such irregular boundaries are also more esthetically pleasing to the observer of the harvested forest.

Pine-hardwood forests may produce abundant food for white-tailed deer. A tract's carrying capacity (the number of animals an area will support with food and water) depends upon the quantity and quality of the browse. The amount of food, and consequently the number of healthy animals, depends upon how often stand thinnings are made, how harvest openings are spaced in the forest, how

Habitat for deer in a hardwood forest. Note that a browse line has begun to form: overstocking is in the offing, *U.S. Forest Service photo.*

large those openings are and how dense is the timber before and after cutting. Fire also plays a role in both providing and temporarily eliminating food plants.

Landowners can exert considerable control over tree stocking during regeneration and harvesting phases of the rotation. Wide spacing at regeneration, for instance, is an advantage for deer, for then the period of maximum animal production is extended several years while the slowly developing young trees form an overhead canopy.

That canopy is important because many understory browse plants produce fruit that deer eat readily. Fruit yields vary according to age and species but—as with browse forage—they are highest in the open. In one case, fruit yields of mature browse shrubs in forest openings was more than 30 times that under pine stands of moderate density. Hence, timber thinnings, made at five-year intervals, continuously stimulate production of browse and understory fruits needed by deer and other wildlife species.

Certain food plants are available for browse only during a part of the year. Habitat management calls for balancing food availability among the seasons.

Regeneration harvests on many small clearcuts, rather than on a large clearcut, add new and usable areas of high-quality food production at frequent intervals. Thirty-acre cutting units provide a practical compromise between larger, more economically efficient harvests desired by industrial purchasers of timber and the smaller areas most pleasing to passersby. Such harvests can accommodate modern machinery and, at the same time, provide enough forest edge for a diversity of food and cover within the home range of deer.

Prescribed burning, as a tool of the forester, also heats up the rhetoric. Burning, indeed, temporarily pollutes the air. A favorable consequence, however, is the improvement in game habitat. Fire kills the tops of browse plants, causing them to resprout near the ground where wildlife can readily forage on these shrubs. New growth is also of higher nutrient quality, as much of the ash of the fire-consumed vegetation consists of chemical elements necessary for plant growth. These nutrients, immediately released into the soil, are used in concentrated form by the new succulent plants. Deer, like cattle, prefer to graze the forage (they are not totally browsers) in forests that have been burned.

With some understory species, like the good winter forage called two-eyed berry, scarcity occurs in openings and abundance in well-stocked stands of forest trees. This is in contrast to yaupon, disaffectionately known as *Ilex vomitoria* to man, but a favorite of deer. While the browse is more important than the seeds, the plant yields 200 lb of berries/acre when growing in the open. None develop when yaupon grows underneath larger trees. Careful management provides for both of these deer foods in the same forest at the same time.

As deer repopulate American forests because of better law enforcement (paid for by hunter taxes) and more favorable habitat, the percentage of those with trophy racks appears to be decreasing. Lack of an adequate diet, of which protein is an important component, is part of the problem. Wildlife biologists no longer attribute poor rack form to genetic inheritance. Some maintain rack size depends upon the condition of the habitat.

The protein matter can be handled by providing supplements in strategically located feeders. But one should not expect to find the trophy buck at the feeder: he comes only after sundown, making several short visits in the night. All the while he is especially leery of danger. Supplemental feeding with high-protein energy feed is most useful in summer when bucks are putting on maximum antler growth.

Although deer with spiked, goat-like prongs, have tasty meat, hunters prize bucks with many-forked antlers arranged symmetrically. Some deer ranges in the southern conifer-hardwood forests are heavily populated with these inferior animals. They should be culled from the herd. Surplus does should be harvested along with the spike bucks to spread out the food supply for animals of both sexes. Removing them enhances the proportion of older male deer with well-formed racks for potential harvest in another year. Poorly formed racks indicate overpopulation for the carrying capacity of the site. Of course, the problem also arises because young bucks are killed before they mature.

One other word: spike bucks are illegal in some states under some conditions. Know the rules before taking one.■

CHAPTER 40

WILDLIFE PRODUCTION— GAME BIRDS AND SONGBIRDS

Woodlots are prime habitat for many kinds of game birds and songbirds. To encourage population increases of many species is a proper and profitable intention for the owner of the land. Managed forests can be good breeding and feeding grounds for a variety of life that will bring pleasure to sports-hunter and bird-watcher alike.

The care of rare and endangered species is a responsibility of the landowner. Sometimes the type of forest harvest will be determined by the present threat to animals and plants. It is too late to save the ivory-billed woodpecker; its habitat required 500 acres of virgin mixed southern pine, baldcypress, and hardwood forest in an unbroken block. We mention the ivory-bill because the pileated woodpecker, a small edition of the former, is sometimes mistaken for it. The pileated (pronounced pi-le-ate-ed) is a regular inhabitant of the forests of the southern Coastal Plain.

Considered here are four other birds, the occurrence of which depends in part upon forest management. They are the American woodcock, the bobwhite quail, the wild turkey, and the red-cockaded woodpecker. To these could be added sparrows, warblers, and many other non-game birds found in abundance among the annual and perennial flowers of clearcut openings and partially harvested woodlands.

A bird not often sought by hunters is the American woodcock. As its habits become better known, landowners may anticipate greater use of the southern forests for the exciting hunt for this quarry.

Woodcocks winter in dense brush cover, called a "picket fence" by those whose trousers are shredded in the attempt to locate this favorite of the sportsman's dinner table. But under the brush must be an opening about a foot high above the ground. The brush obscures the woodcock from overhead predators while the cleared ground enables its escape from surface enemies. Such brushy sites are frequently fertile soils of a loam mixture and with good water relations. Sites that have been burned or that

The pileated woodpecker, the largest North American woodpecker (assuming the ivory-billed is extinct) carves a nest in a dead hardwood tree. Bird books may say the pileated vacates a cutover forest, never to return. However, like many forms of wildlife, this bird acclimates to man's intrusion as long as suitably sized snags and some mature trees are left for them. A pair regularly visits the author's yard in a small city, *Richard Conner photo.*

are naturally high in calcium make especially good habitat, for here earthworms—a favorite food of the bird—are most abundant.

Forest game managers are able to create favorable habitat for this game bird with carefully prescribed fire and controlled grazing of cattle. To retain the flocks after crowns close in pine plantations, needleleaf species are periodically thinned and broadleaf trees killed. While the hardwood sprouts that arise after the thinning and deadening compete with the conifers for soil moisture and nutrients, the broadleaf shoots become the picket fence, the vertical vegetation that is so important as habitat for the bird.

The thrill of a lifetime came for one young hunter on the first outing with his new Brittany spaniel. An American woodcock was the first bird pointed. It was also the first of the species the youth had ever seen, and it was taken with the first shot. The bird, wintering in Texas, had a legband labeled Wisconsin.

Earlier times saw the cutting of forests for agricultural pursuits as well as for lumber. The openings made by this high-grading harvest, when lumbermen did not think about future stands of timber, resulted in a patchwork quilt of logged sites, burned areas, grazed lands and plowed fields. The variety of vegetative cover that followed the harvest made possible an abundance of upland game animals and, among them, the bobwhite quail.

By the 1940s, a team of forest game managers in the Southeast began recommending fire for improving the habitat, and therefore the numbers, of bobwhite quail in the pineries of the Coastal Plain. While their early work showed the need for prescribed burning in the longleaf pine forests, later studies alerted managers to the usefulness of this forestry tool in other timber types. Fire is a lot cheaper than plowing the soil to prepare an area for game-food plots.

After the site is prepared, landowners should plant cereal grains and woody cover in these openings in the woods. Later the cultivated cover crop is fertilized with commercially available nutrient amendments.

In the southern coastal plains, quail habitats are improved by maintaining vigorous stands of longleaf pines. To do this usually requires prescribed burning for seedbed preparation, needle-disease control, and hazard reduction. The procedure encourages the natural introduction of quail food plants.

Maximizing quail populations while maintaining adequate forests of pine requires cool, fast-moving fires. These prescribed burns consume the vegetational debris and thereby open the ground for germination of seeds that produce food essential for the birds. The carrying capacity of the site for bobwhite quail is, as a result, greatly enhanced.

Hardwood trees will not be eliminated by a cool fire; they will only be retarded. The mast-producing oaks and hickories, along with associated species like persimmon and sassafras, are woodland stages in ecologic succession of habitat development. These are desirable for bobwhite quail covey protection. Some of the best bird hunting in the world is in such forests.

Wild turkeys, whose metallic gobble is now heard in many American forests where it had been hunted to extinction, appear mostly in river bottoms. The bird may thrive, too, in upland pine-hardwood forests, where it may never have been native. The "Pilgrim's pride" fell on hard times when the virgin forests were harvested in the years of cut-out-and-get-out economic policy. "Planting" pairs of birds moved from higher-density areas, coupled with law enforcement that precludes poaching, has almost doubled the turkey population in the South in recent years. From the few scattered flocks remaining 30 years ago, the population now permits legal harvests of over 60,000 a year.

For the majestic turkeys, happily, most modern forest management practices enhance their vigor sometime during the bird's life cycle—nesting, feeding, growing to maturity and reproduction. Supplemental feeding utilizes strips, disked through the woods and planted with millet, a favorite of the large bird. Firebreaks plowed and planted to rye and oats provide winter food. Hunters should note, however, that some states prohibit shooting game over baited fields. Whether a planted field is "baited" may be a matter for a judge to decide.

As on the range and in the garden, soil pH, lime, nutrients, and texture play a role in game-bird production. Partridge peas, annual legumes that are dessert for birds, come in various varieties and species. The most robust for a particular site as a supplemental feed depends upon the character of the soil.

The red-cockaded woodpecker began to be

missed in the southern pineries as fire prevention encouraged dense growth of understory brush and vines. To accommodate the bird, prescribed fires are encouraged in order to reduce the density of the woody vegetation hugging the ground. This improves colony sites. These woodpeckers make their houses in hollowed-out boles of living pines, where one nesting pair and several helpers form a colony. The resinous sap from the wounds they make in carving out the hole often coats the bark of nest trees. If this sap is ignited, the flame may climb the tree, burn its crown, and destroy the bird's residence. Burning must be carefully prescribed and loose pine straw and dead leaves raked for a distance of about 10 feet from around the bases of nest trees that visibly display resin flow. In colony sites, the trees should be safeguarded by plowing a furrowed fire lane around the area.

Potential activity sites can also be destroyed unintentionally by timber-markers preparing a site for thinning. Trees with red heartrot infection, the living boles in which the woodpecker carves out its nest, are usually those selected for harvest. Their removal enhances the residual stems that have been left to grow to add high-value wood for a future harvest. But without these nest trees in the woods, red-cockaded woodpeckers could again be threatened to extinction.

Discovery of the ecologic relationship between the bird, the tree, and the disease suggests that a red heartrot-infected tree should be retained on every 20 acres. Of course, the infection spreads to introduce rot into healthy stems. Whether to maximize timber production or to cater to threatened wildlife is a decision for the landowner.■

C H A P T E R 4 1

WILDLIFE PRODUCTION— FOR SQUIRRELS

The future for squirrels is not as promising as desirable in much of North America. In fact, some major and costly measures may be needed if these rodents, because of their affinity for certain timber types, are to continue as the most-hunted game in the U.S. The problem also relates to the nation's shrinking bottomland acreage that remains in forest. Ranches, farms and lakes take much. That acreage reduction necessitates growing more wood on the remaining tracts. And if that wood is in conifers (the most important type for construction and paper), less oak and hickory mast will be available to supply food for squirrels.

Squirrel-produced bark fibers at a nest opening in an oak provide a clue that the forest is stocked with the small-game mammal, *Chester E. Billie, Jr., photo.*

Cull trees for squirrel dens and mast must be left in silvicultural operations. Doing so, of course, reduces future commercial timber-growing potential on the space taken by these stems. At least one such tree on every acre should be retained in any thinning or harvest.

Gray squirrels roam among bottomland hardwoods while fox squirrels inhabit pine-hardwood types, only rarely venturing into the bottoms. Both prefer nut foods. Some squirrels feed on conifers: the endangered Kaibab squirrels of the Arizona Grand Canyon country in the winter eat almost exclusively the cambium, the layer just beneath the bark, of yellow pines like ponderosa. Fire prevention in stands of these trees in the plateaus of the southwestern U.S. may have inadvertantly caused a decline in this and other similarly feeding squirrel populations. Without fire, necessary for preparing the site for regenerating these yellow pine stands, few young stems take the place of old trees as they die. The trees would not have been suppressed had fires been allowed to thin them. Accessible cambial tissue is then not readily available. Squirrel starvation results.

Fox squirrels in the East seem also threatened. Here the demise may be directly related to habitat changes brought about by timber felling, the broadleaf virgin forests replaced frequently by coniferous stems which today are more economically important. To protect small mammals, the landowner may need to give up maximizing income from timber production.

As a rule, ground squirrels favor herbaceous growth for their habitat, joining pocket gophers and shrews in the new vegetation arising after clearcutting and burning. They remain on the site, joining the rabbits and chipmunks, as tree seedlings and saplings and grasses intrude on the land. In contrast, flying squirrels and red squirrels inhabit older, mature stands without natural openings and where there is little herbaceous vegetation.

As a general rule, cutting half of the basal area of a forest has no effect upon survival, reproduction or density of gray squirrels. Greater removal, especially clearcutting, temporarily diminishes their numbers. Selection harvests are thus encouraged. If clearcutting must be done, the openings should be less than 10 acres.

Landowners may need to consider opportunity costs to estimate the financial value of wildlife. Here is an example of the procedure for a squirrel: a 60-year-old bottomland oak-hickory woodland contains 12,000 bd ft/acre, worth $40/MBF on the stump. That timber standing in the woods is worth $480 an acre. The owner may choose to retain the trees, foregoing the opportunity to receive $480 without any effort, in order to provide habitat for the small game mammals. Or the owner may elect to harvest the timber. The $480 spread over the 60-year life of the stand (without considering past costs and interest rates) gives an annual value of $8.90. On a single acre, most owners would choose to cater to the squirrels. But if 10, 50, or 200 acres are involved, the opportunity facing the landowner becomes more acute.

Now, suppose the carrying capacity for squirrels is one/acre on this tract. Or two, or 10 of the rodents per acre. If the land is left in trees only to accommodate the animals, how many squirrels it will support plays a role in the management decision. If the rodent has a monetary value of $20 per head, it must be readily marketable at that price while running wild in the woods. One such animal on each acre, if it is killed by a hunter willing to pay that price, suggests to the owner the desirability to retain the broadleaf mast-producing woodland rather than to harvest it. On the other hand, a conservation-minded landowner may logically reason that, while he does a disservice to squirrels in harvesting his broadleaf timber for furniture and panelling woods, he provides for the needs of people.

Owners of raped forested land may anticipate rehabilitation of such tracts for future squirrel hunting. An owner may find he has been handed the wheel of a doomed ship. Nature, however, is a great rehabilitator. Depending upon the site, its quality for producing squirrel-habitat vegetation, and the availability of seed sources, a decade or two may find new families of this small game animal populating the land.■

C H A P T E R 4 2

MANAGING FOR RECREATION

These poetic lines of Ralph Waldo Emerson, penned in the mid-1800s, describe the attitude of many Americans toward the forests of the small landowner:

I do not count the hours I spend
In wondering by the sea.
The forest is my loyal friend,
Like God, it uses me.
Aloft in secret veins of air
Blows the sweet breath of song;
O, few to scale these uplands dare,
Though they to all belong.

Maybe a century ago few ventured "to scale these uplands" in the forest. But all people—even those who did not hold title to, or pay taxes on, the land—with impunity hiked, fished, and hunted on it. The forest was open for all to enjoy. Indeed, the Congressional Act that set aside Yellowstone as the first national park designated it "a public pleasuring ground."

Increases in population and the ease of transportation changed that. Greater numbers of citizens for a total land area that does not increase in acreage raises the value of each acre. This is particularly so for woodlands, especially if the forested tract is near a population center. More leisure for the working classes, along with greater mobility because of good roads and inexpensive personal transportation, hastened the change.

Poaching that occurred on the land gave legislators the impetus to pass hunting laws. Horses and cattle that once roamed unfenced ran into conflict with automobiles and trucks, forcing lawmakers to enact statutes that require confinement of large animals behind fences. Vandalism by uninvited visitors on private property hastened the need for property-protection statutes. With punishment prescribed for trespass (the legal term for illegal entry), landowners next posted their lands to exclude passage for any purpose. In some areas, posting backfired: on lands so marked, fences were cut and fires set.

Meanwhile, to accommodate landowners threatened with exorbitant liability insurance, state legis-

Many families use a woodland for recreation as the trend for a simpler life-style develops.

latures provided release from such litigation brought by anyone injured on the property of another, whether they had ventured by invitation or as trespassers.

Uninvited guests injured on the forest lands—referred to in legal jargon as attractive nuisances—of another needed to be dealt with by appropriate laws. To further restrict the trespass of these people, landowners strengthened their fences, chained and padlocked gates, and hired "fence-riders." Once these were the pasture-riding horsemen on the western range searching the boundaries of extensive tracts for stray cattle and rustlers. Now, fence-riders use four-wheel-drive vehicles to check the interiors as well as the perimeters of large and small tracts for all kinds of trespassers and their activities.

Add to the changes in land use due to population growth and legislation, changes in society's values. Leasing of land for hunting and leisure resulted. Hunting leases, enabling lessees to take specified game (such as deer, squirrels, woodcock and bobwhite quail) or all kinds of legal game often provide the landowner enough money to pay the property taxes on a tract. These lessees usually explicitly exclude the erection of permanent structures, but may designate a site on which lessees may pitch tents, park camper vehicles and build fires.

Lessors may, and should in contracts, exclude the use of alcoholic beverages while on the property (lest an injured party find reason to sue, in spite of laws and contracts which would *seem* to protect the owner). The lease should include a map and show beginning and termination dates. Game leases may be for less than a calendar year. They also may be for longer periods. Leased land should be well-marked (fences preferred) to prevent lessees from straying on to the property of another.

The minimum-size tract for a game lease in an isolated area is 200 acres, although some are as little as 40 acres. Lease tracts surrounded by, or adjacent to, open lands for hunting (such as national forests and some state forests and parks) could be as little as 10 acres, the site principally serving as a base camp for the hunting lessees. Often, however, these small leases become base camps for poaching.

Ownerships of a thousand acres could be utilized for exotic game-animal production and, when the carrying capacity of any species is reached, leased for the taking of a particular animal species on a particular day. This kind of program becomes highly entrepreneurial: hog-proof fences as high as 4 ft and topped by another 4 ft of barbed wire are necessary to confine the game. If far distant from accommodations, lessors build quarters or provide tents for patrons. State game laws do not apply: exotic wildlife are treated legally as domestic livestock. This philosophy—that such wildlife is not the responsibility of the state—is in the throes of change.

Successful hunts may gross the owner a minimum of $1,000 for each kill, depending upon species, trophy quality and abundance of the game animal taken. Expert hunter-cowboys usually back up the client-hunter to bring down wounded animals, lest they wander off to die. Several introduced sheep species (including barberry and barbados); axis, red, fallow and sika deer; sambar; and blackbuck antelope are now bred on wildlands in the U.S. for this purpose. Ring-necked pheasants, both naturally reproducing in the wild and produced in hatcheries and freed for the hunter's fare, provide exciting game. Landowners should have an ecologic and economic appraisal of the property made by an expert game biologist prior to considering this investment seriously.

As forested properties available for family leisure dwindle due to population pressures and the resulting higher land values, owners of tracts within an hour or two from cities may consider leasing 10- to 100-acre parcels within the tract for hiking, camping, and fishing. Many families confined to the "concrete jungles" of metropolitan areas value highly the opportunity to consider a small tract as their own for a time.

An appropriate annual fee for such a lease might be equal to 3% of the land's value. Thus a 10-acre tract valued at $1,000/acre could return to the landowner $300/year. The owner of a 100-acre woodland could lease out 10 such areas, prices for tracts increasing if they front on a lake or pond or through which a stream runs.

To protect their leased land and trees, owners should designate camper and tent sites, spell out financial damages for injured trees (kids girdling and lanterns scorching them) and soil erosion (trucks bogged down when the ground is too wet for their entry on the land). The tract need not be fenced if painted trees mark the boundaries of each lease.

Lessees will want to get away from others for awhile, but not too far away. Neighbors on adjacent 10-acre tracts provide no significant intrusion. Although they may not be gregarious, most people appreciate the presence of others nearby. Owners should be aware that these user-clients will bring to the woods their urban ways, naive about much of the unfamiliar which they will encounter in the rural setting.

Lease contracts will need to spell out clearly lessee obligations to himself, his family and visitors to the leased site, and to the property. An understanding of health and safety for city dwellers temporarily camping in a pristine environment must be underscored. The falling branch, the windstorm-tossed two-ton tree, axe-made incisions, pit vipers and coral snakes, poisonous plants and the first-time use of a gasoline-mantle lantern pose problems for the unwary.

These tenants usually bring simple and minimal equipment to the site. They will engage primarily in picnicking, hiking, fishing, bird-watching and nature study.

To protect the property-owner's interest, deposits are desirable. Their return may depend totally or proportionately upon user impact. Each year the site should be monitored for evidence of new soil erosion (determined by exposure of roots, rock and subsoil). Owners should maintain a record of the rates of infiltration of water into the surface soil. They should also observe surface soil drainage and tree damage (again, kids using axes on, and driving nails in, living trees and burns from a lantern hung on a limb too close to the bole of a tree). Contracts should describe sanitation requirements, the minimal toilet permissible, and whether lessor or lessee provides it. The legal document should specifically forbid the use of off-road recreational vehicles (ORVs) if the owner values the land.

Fishing shanties along river edges provide pleasure for many.

Lessees of small tracts for family recreation likely will be financially conservative. They will be members of the lower middle class, with time for leisure but not with the money for extravagant or flamboyant gamesmanship or extensive travel. They could also be folks seeking a return to a simpler life-style. With these kinds of user-clients, a landowner's property should be in responsible hands. They could well be people who paraphrase, "The family that prays together stays together" to, "The family that plays together stays together."

Dude ranching and vacation farming may also provide income to landowners while they await a timber harvest. These may be participatory or just watching a ranching and farming operation while simultaneously enjoying the beauty of the owner's woodlands.

Income may also be obtained through conservation easements and covenants. These agreements, generally made between landowners and an agency of government, provide regular payments to the owners by the government. These payments assure the continued and same use of the land. Conservation easements save small woodland parcels from being developed when areas are rezoned due to the pressure of urban population growth. Higher property-value zoning often pushes taxes out of reach for forest owners, forcing them to sell woodlands to developers. Conservation easements rescue the lands from this potentially perennial threat.

"Greenbelts" around cities frequently consist of these agreements. Woodland owners desiring to retain a tract forever in its unimpaired condition should recognize, however, that if people do not manage the land, nature has a way of doing it for them. Various insect depredations in North America attest to this.

Some state governments also lease privately

owned tracts for short-term wildlife refuges. These temporary protection areas are retained until the carrying capacity of the parcel for the particular species is reached, or nearly so. The owner then receives the land back to be hunted according to game laws and regulations prescribed by the authorities for the parcel.

Tax breaks are available from many local and state governments for leaving a woodland in its natural state. The cost in lost timber due to a lack of management may exceed the benefit to the owner of the lesser tax.

To aid a landowner in deciding the current value of land when such a contract is considered, this formula comes in handy:

$$\text{Value} = \text{Annual income/interest.}$$

Thus, for a tract which has been producing an average of $1,000/year (not adjusted for inflation) over the past 20 years, and the owner considers money worth 5%/year, the value of the property for this purpose is $1,000/.05 = $20,000.

Owners of small forested properties should seek assistance for local situations from the Agricultural Stabilization and Conservation Service (ASCS), the Soil Conservation Service (SCS), and the Farmer's Home Administration (FHA). Funds may be available from the federal treasury through the 1962 and 1965 Congressional Food and Agriculture Acts under the Rural Area Development Program for small landowners interested in outdoor recreation opportunities. (The work of the Heritage Conservation and Recreation Service is now handled, for the most part, by the National Park Service in the U.S. Department of the Interior. Its functions ordinarily do not include the concerns of the landowner of small forested tracts.)

In all of these uses of woodlands, the owners should anticipate intensive management of the timber. For shelterwood and selection harvests (Chapters 7 and 8), tree removal should have little impact on recreational leases. Clearcutting and seed-tree regeneration (Chapters 5 and 6) discourage leisure leasing. The latter methods, however, will improve hunting opportunities for some game animals, depending upon forest cover type, game species, tree age and size of openings.

Beginning this chapter with the rhyme of one New Englander, it seems appropriate to conclude with another. Robert Frost's oft-quoted lines:

Whose woods these are, I think I know.
His house is in the village though;
He will not see me stopping here
To watch his woods fill up with snow.

The poet continues, The "woods are lovely, dark and deep." That description for many is enough.■

C H A P T E R 4 3

CONSERVING THE SOIL FROM WIND EROSION

Windbreaks, long used to reduce the movement of soil, were first introduced in the United States in 1903. At that time, 16,000 acres were planted with ponderosa pines on the Nebraska National Forest, in those days referred to as a forest reserve. Although this stand of trees was not arranged in rows, its purpose was to minimize wind erosion.

Trees, serving as windbreaks and shelterbelts, offer a useful technique for stabilizing soil, especially in the plains and prairies of the American West. Many of the techniques for developing these protective barriers had been earlier employed in England, France, and the steppes of Russia. The hedgerows made famous by Winston Churchill's emotional "We will fight" speech and the boundary vegetation encountered by soldiers in the battlefields of Normandy had been planted to obstruct the wind.

A tale is told of Franklin D. Roosevelt traveling the Great Plains in the Dust Bowl days. The President's train encountered troubles as the blowing soil accumulated along the rails. It was quite natural for the President, who signed his income tax forms as a forester, to propose that the government plant rows of trees from the Canadian border to the Rio Grande to protect the soil from scouring winds.

On hearing the suggestion, a White House staffman sent for Raphael Zon, head of the U.S. Forest Service's research division. Ushered into the President's railroad car as the train tunneled through blowing silt, Zon and Roosevelt laid the groundwork for the shelterbelt and windbreak programs of the Great Depression years.

Isolated rows of trees, as some say the President had in mind, were supplanted by Zon's rows of trees grouped together. Sometimes three rows spaced 10 ft apart, more often five rows or seven, are planted. Each row is of a different species. Later these trees will be expected to provide food for game, wood for fuel, Christmas trees, and lumber for furniture and farm use. Hence, such trees as holly for quail food, oak for fuel, thuja for yule trees, walnut for tables and red pine for construction were planted in the plains and prairies.

Fortunately for Americans, Zon, as a Russian immigrant, had worked in soil erosion control in the steppes of that land. Fleeing the new communist government, he brought his knowledge, education and experience to the New World, probably never dreaming how—during a time of drought and bank-closing Depression—the information he carried in his head would save us years of testing and trials in shelterbelt management. The Russians were ahead of us in soil conservation practices.

Shortly, foresters would venture into the communities of treeless prairies and plains to assist farmers in species selection, ground preparation, planting

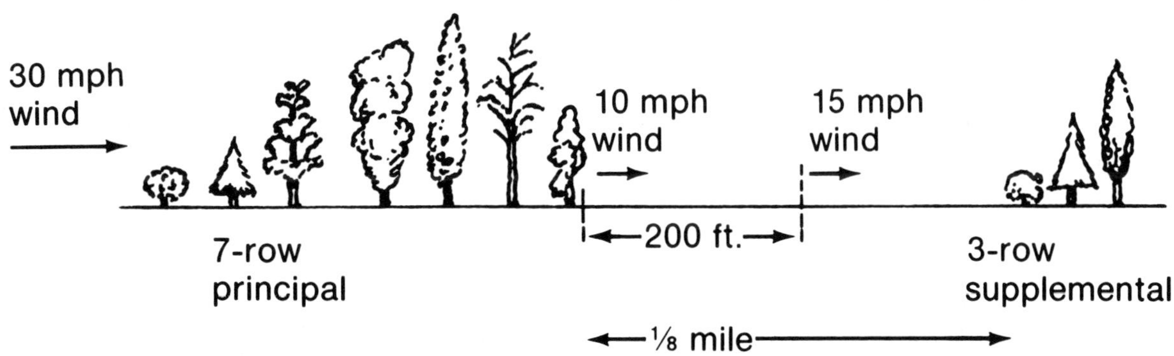

Seven-row and 3-row shelterbelts of various drought- and wind-hardy trees protect soil from wind erosion.

techniques and, later, silvicultural practices such as pruning and thinning. Still later, according to the plan, they were to guide the harvest and sale of mature trees.

Two federal programs, both in the U.S. Department of Agriculture, were initiated at about the same time. The Soil Conservation Service was assigned *windbreak* plantings, principally for around rural homes, barns and pastures. Not only would soil be conserved, but reduction in the speed of chilling wind would keep homes and barns warmer in winter and, by providing some protection for livestock, reduce the amount of feed they require.

Depending on regional climate, windbreaks also protect orchards from drying winds and, important in these days of high energy prices, reduce fuel consumption in homes by as much as 25%. In the summer, shade tree canopies lower room temperatures by as much as 20 F. Mulch provided by the foliage reduces evaporation loss from the soil and, upon decomposition, incorporates additional organic matter into the surface soil. That added humus improves retention of soil moisture and encourages rapid infiltration of rainfall into the surface horizon.

Windbreak remnants remain, still protecting windward—usually western and northern—sides of structures and pastures. In the absence of these stands, wind-blown sand, silt and clay often fill drainage ditches and irrigation canals, bury fences, cause traffic hazards and damage the health of man and livestock.

The Forest Service had its program, too. Called *shelterbelts*, these protective stands need not be located near residences and barns and, indeed, more often were planted remote from habitation in rows generally running north and south in order to reduce the westerly prevailing winds. A few shelterbelts remain as reminders of a noble effort to contain the wind and retain the soil. That noble effort was carried out at a time when many farm folks were leaving their homes.

Beaten by the wind, some went to California. Others migrated to the eastern cities, only to find that the wind-driven dust had settled on the metropolitan areas there. Temperature inversions—the same "upside-down climate" air phenomenon that causes smog to hover over congested locales—force the fine particles of once-rich soils to descend. Some city housekeepers of those years, chasing the persistent dust on the furniture, knew the situation about as well as today's farm wife of the western plains.

Landowners are dealt a devastating blow by nature when high winds accompany drought and no vegetation covers the land. The two phenomena ordinarily account for a loss of a billion tons of soil a year in the U.S. That is about one-fourth of the weight of soil lost by water erosion from North American lands to streams, lakes and oceans. In extraordinary droughts, wind erosion is even more severe than that caused by water.

Some soil, of course, can be lost without greatly reducing productivity. Indeed, erosion is a phenomenon of nature and, literally, is "as old as the hills." On deep soils, an annual loss of as much as 5 tons/acre can be tolerated without greatly hurting production. Shallow soils are damaged seriously by much lower losses.

Wind damage to land in the Great Plains in 1977 was the fifth worst since federal records began in 1935. Damage to 2.1 million acres of land during a six-month period in 1976-77 occurred in Texas, the worst there in 20 years. That compared with 1.5 million acres the year before.

The Soil Conservation Service's (SCS) measure of damage may over-dramatize the problem; but even so, the problem is one for which a solution is, in part, available to landowners. SCS tallies land as damaged if enough soil is removed from a given tract to intensify later erosion on that site. Land is also considered damaged if soil accumulation around or over crop plants and range grass is deep enough to reduce their growth. Hence, soil blown from one field to another may reduce current productivity of the land from which the soil is moved. Then, in later years, the deeper layer of silt or clay in the field on the lee side enhances growth there.

A recent U.S. Congressional report said 84% of western American farmers are losing more than five tons of soil/acre each year. While much of that loss is not attributed to wind, controlling wind intensity will help where soil losses hurt most. As new topsoil forms by freezing and thawing and the dissolution of rocks at a rate of about 1½ tons/year, losses to erosion of more than this amount in time deplete the land. Substituting fertilizer, irrigation water and pesticides for vanished soil are temporary, but expensive, remedies.

The effectiveness of trees in holding down wind speeds and, hence, checking soil movement, is readily measured. A 30-mph wind passing over a seven-row shelterbelt is reduced to 10 mph on the lee side. It increases again to 15 mph at a distance of 200 ft. At that distance, supplemental rows of trees (usually three rows spaced perhaps 1/8 mile apart) are planted.

Some citizens would have Congress require that landowners participating in U.S.D.A. financial assistance programs be first required to follow specified soil conservation practices. Thus, to be eligible for support, crop insurance, or disaster relief, landowners might be required to show that they have planted and cared for their windbreaks or shelterbelts.

Proponents contend that the mid-1970s dust storms showed that soil conservation necessary to sustain agricultural productivity of the Great Plains and other farming regions had not been achieved. What had happened to the windbreaks and shelterbelts so painstakingly lining the land in picturesque rows in the treeless plains?

Landowners through the 1950s saw the acreage in trees as an encumbrance in an industry which all too often operates marginally. A mile of shelterbelt easily utilizes 10 acres of high-yield land, soil needed for agricultural crop production. The reduction in crop production on either side of the belt may "cost" the equivalent of eight more acres for each linear mile of trees.

New center-pivot irrigation systems are encumbered by windbreaks and, if ground water is plentiful and available, reduce the need for the trees. Moist soil is not likely to blow. Still another cause for shelterbelt removal is the poor maintenance of the stands: disease, dead trees, and stem breakage leave the rows unattractive and unable to accomplish their aim.

Weather is cyclic and man's memory frail. The rains returned and for most of three decades the great blowouts of the drought years ceased. But faith is fragile, all too easily shattered and lost. So too is memory. Just as the drought of the Great Depression era (what had we done to deserve both drought and economic disaster at the same time!) was followed by adequate rainfall, another season of moisture stress began in the 1970s. Though many of the planted rows of trees are gone, Zon's techniques remain ready to be put into practice. (The Forest Service maintains, at Lincoln, Nebraska, a research center specializing in how best to quickly establish plantings to protect the soil.)

For much of America's mid-section and West, wind is more important than water in causing soil erosion. This includes the vast area where high winds are associated with low rainfall. Without enough moisture to bind particles of soil into aggregates that resist disturbance by wind, the fine particles sail great distances. For example, soil from West Texas was reported over peninsular Florida some days after the early-1977 dust storm. Wind effects are so calamitous that a shelterbelt may be paid for by saving a single year's crop.

Shelterbelt care, and renovation of those in poor condition, should encourage their retention and, hence, improve crop yields. Care begins with planting. Sandy and loamy soils are recommended, meeting adequately the moisture and aeration requirements of tree species to be used. Deep soils with a stable water table are preferred.

A typical plan calls for a seven-row shelterbelt, likely consisting of seven species of shrubs and trees. From the windward to leeward side, the first is a shrub, perhaps a hedge species, that hugs the ground even after tree crowns in adjacent rows are well above the surface of the land. Next, a conifer, like pine, with evergreen foliage, provides protection year-round. A deciduous hardwood of medium growth rate and a row of another species with faster growth, perhaps Chinese elm and catalpa, respectively, follow in the center of the belt. Then toward the lee side, a fast-growing short-lived hardwood, like cottonwood, and a slower-growing high-quality wood, like black walnut, are appropriate. Finally, a holly, an evergreen broadleaf tree, completes a belt. It provides food for wildlife.

Tree seedlings for protecting the site from wind are conveniently planted with mattocks or sharpshooter spades. Seedlings must be kept moist. Even a half-minute of root exposure to drying winds while removing the tree from the carrying pail, inserting it in the hole, and filling the hole can reduce the seedlings' vigor.

All of the root must be in the ground; any exposure will likely kill the tree. One cannot plant too deeply, as long as the terminal bud is above the ground. Placing the tree so that the root collar is an inch or two below the ground level assures against later erosion, when wind blows or rain washes away some soil and, thereby, exposes roots.

States make available many species generally

adapted to the climate of the region for windbreak planting. Conifers grown to protect soils, homes, barns and grazing cattle from the cold and drying effects of wind include Arizona cypress, a beautifully conical native of the U.S. Southwest; Austrian pine, an import noted for its pleasing shape and stiff needles; and southern baldcypress of swamp fame. It is a tree that grows well on moderately moist sites if competition is controlled.

Oriental arborvitae, known to many as cedar or thuja; ponderosa pine, the picturesque tree growing in park-like stands in America's West; and eastern redcedar, the juniper commonly spread by birds along fences and phone lines, are also planted. Loblolly, shortleaf, slash and Virginia pines, common to the South, are used in windbreak plantings.

Forest tree nurseries provide some broadleaf trees. Black walnut, a potentially valuable furniture wood; catalpa, of fishbait fame; the exotic Chinese elm; and cottonwood, a rapid-growing stem that sprouts from 12-inch cuttings inserted in the ground, are available. Other species which may also be purchased for protecting soil from wind damage

A windbreak of cottonwood, planted as cuttings, protects this home from energy-consuming wind.

are honey locust, making a good fence post even when untreated with wood preservatives; osage orange (sometimes called horse apple or *bois d'arc*, for its importance to the Indian for hunter bow manufacture); sweetgum of the famed-grain (redgum color); sycamore; and bird-loved mulberry.

Many kinds are exotics, brought from abroad for the purpose. For example, eucalyptus, an introduction from Australia and highly susceptible to cold weather damage, is planted in the warmer subtropical zones of the country. Some, like Russian olive, may be associated with the immigrant from that land who had so much to do with successful establishment of shelterbelts in the U.S.

A minimum of 500 trees usually must be purchased. Trees are ordinarily shipped collect or picked up at nurseries. They are packed usually in bales of 1,000 seedlings. Shipments begin after the first frost and continue until March unless stock is exhausted earlier.

Seedlings are packed for shipment in wet sphagnum moss or a synthetic substitute (like Terrasorb) to keep the roots moist and bundled in waterproof paper. Orders should indicate the planned planting time and arrangements made to do the job shortly after receiving the trees. Seedlings can be left wrapped a few days if kept cool and moist, but their survival is best when planted promptly.

Competing vegetation must be controlled. Preparing the site by allowing the land to lie fallow for one or more growing seasons before tree-planting helps store adequate water for the seedlings' first season in the field. It is good for the soil to be moist to a depth of at least 3 ft at planting time.

Cultivation of the land in such a way as to retain an organic mulch is important. This residue lessens water losses through evaporation by increasing the rate at which rainfall infiltrates into the soil surface. Chemical herbicides that are pre-emergence (kills germinating weeds), long-persistence (effect extending through one or two seasons) and selective (will not injure trees) may be used where machine cultivation is impractical. In the interim between plowing and tree-planting, the soil may be protected by an annual winter cover crop; later the trees are planted between crop rows.

After planting, weed control is necessary until tree crowns close; the wider the spacing, the longer the period for vegetative competition. For mechanical weed control, a shovel-type cultivator or "grape-type" hoe is recommended. To avoid cutting surface-hugging roots of windbreak trees, cultivation should be shallow. It also should be frequent, killing weeds while they are small. Since tandem discs pulverize the soil, they should not be used. Proper cultivation produces soil clods; these are less likely to puddle during rain or to blow when dry.

Organic mulches may be recommended for moisture retention on these lands. In some sections, however, mulch may harbor tree-injuring rodents which chew through seedlings and girdle saplings. Organic mulch may tie up nitrogen, used by bacteria for natural decomposition, which otherwise would be available for tree growth.

Sheets of black plastic may substitute for organic mulch. Cut square, 4 ft on a side with a hole in the center for the seedling, these effectively reduce evaporation and warm the soil above normal temperatures. Weed growth is also checked by the plastic sheet, held down at the corners by a shovel-full of soil.

Insects—especially bark borers and foliage chewers—take their toll on shelterbelts as they do on all forests. Hundreds of such pests plague the Plains. The forester's complaint that insects feed on poorly managed stands—those of poor vigor—appears true for windbreaks of the western flatlands. Insect-resistant strains of tree species are sought; but as yet, the search has not been successful. Pesticides are available for the important nuisance insects.

Diseases infecting these tree plantings include cankers that deform trunks. Trees are then easily broken by the wind. Stem rots, which cause wood decay, root rots that encourage windthrow and foliage-destroying blights and rusts are other fungi problems. The expense of control ordinarily precludes the use of fungicides, though chemicals are available for some of these maladies.

To rehabilitate old windbreaks, landowners resort to thinning that removes weakened, unhealthy stems. Good trees, if crowns are completely intertwined, are also cut. Ordinarily, interplanting of younger trees or seedlings in the "skips" is discouraged. These are unlikely to be able to compete with the overtopping vegetation for soil moisture, nutrients, and light. Adding whole new rows of trees, however, may aid rejuvenation.

Pruning dead or dying branches and those with cankers is appropriate. A tractor attachment for this purpose is available. Fencing to exclude cattle could be desirable. Irrigation, too, may be recommended if water is readily available and if farm crops, under the same circumstances of weather and soil, would benefit.

The cost of a windbreak in relation to its benefit for controlling wind erosion varies by region. Studies to chart the saving of soil and other benefits—such as improved crop yields, landscape beauty, wood products, livestock protection, and wildlife habitat—are available from the SCS. That bureau has primary responsibility for windbreak and shelterbelt work on small ownerships, including management and, under certain conditions, cost-sharing.

Trees for wind protection often have been a novelty. Likely they will become a necessity for good stewardship of the land, especially in the windy plains and prairies.■

C H A P T E R 4 4

LANDOWNER'S SHADE TREES

The recent past has seen a great upswing in interest in shade trees and their care. Cities hire foresters to look after their trees, power companies publish elaborate manuals about them for distribution to customers and Congress earmarks a sizable portion of the Forest Service's budget to urban, or shade-tree, forestry. Thus, planting the right landscape tree in the right place is an appropriate concern for achieving pleasure today and enjoyment for years to come.

Some house builders are landscape wreckers. An esthetic natural woodland is converted into a desert-like red clay plain by cutting trees and filling, digging and packing the soil. Gullies develop as the precious little surface soil left washes down to muddy up streams.

To this scene, many new rural (and urban) homeowners despair, wishing for a Johnny Appleseed who, with 20th-century vim, could reforest their yards with tall shade trees. Those unable to afford transplants of pole-size trees need not be discouraged. They can be their own Johnny Appleseed. And they can get results faster than did the Ohio settlers along the route of Johnny's journeys. In a handful of years, children can romp under the shade of yellow-poplar, elm, oak and ash. These paragraphs outline only some broad generalities about trees to plant around the rural home.

Depending on the climate of the region, winter (if the ground is not frozen) to spring (before buds break dormancy) is the time to plant. A wide variety of rapid-growing forest species for yard improvement are commercially available. The landowner's woods will likely have many useful species. With a little extra water and annual fertilizer applications, growth is stimulated appreciably. Since water from America's rivers is expensive by the time it reaches a yard faucet, its use may be restricted to dire droughts—perhaps periods of three weeks without rain.

Home landscaping attitudes vary, for "beauty is in the eye of the beholder." One may wish his yard to appear as a natural woods, planting many species of trees in a random manner. Another wishes the house to be framed with a tree at either end, much as the face of a girl is bordered with a page-boy hair-style. Some may even prefer the "busy" look, a dense stand of high and low vegetation, thickly intertwined, that hides the home from view; while another may prefer the orderly rows of a plantation, as though laid out by an engineering surveyor.

Trees may be planted for quick landscaping of a "desert" property (try hybrid poplar or cottonwood cuttings), for shade (an elm) and for color (dogwood in the spring and red maple in the fall). For

A weeping willow for a homeowner's landscaping. Beware, its roots grow into the seepage of a leaking sewer and, thence, into the pipe to clog it.

esthetics, a multiple-trunked tree (like eastern redbud), a conical-crowned one (pin oak), or large spreading branches on a short-bodied bole (live oak) may be preferred. To attract birds, one plants an American holly; for squirrels, a hickory; and for deer, the yaupon of American Indian stomach-cleaning fame. (Its scientific name is *Ilex vomitoria.*)

Sycamore and Chinese elm respond well on poor soils. While the former quickly produces a stem and roots from sprout cuttings inserted in the ground, seedlings generally survive and grow better. Kinds of trees which respond more slowly, but which do satisfactorily on sheet-eroded sites when fertilized, include southern magnolia, Chinese chestnut (a much-forked relative of the famous American chestnut with inferior-tasting fruit), most oaks, hemlocks (in the shade), slash pine, Virginia pine and dogwood.

Trees to plant on better land—where the earth is deep, dark, moist, and loamy—include white pine, Arizona cypress, maple and shortleaf pine. Although the latter grows naturally on extremely poor sites, frequently it is killed on such lands after 20 to 30 years by the littleleaf disease. Where shortleaf pines have acquired this malady, heavy annual doses of nitrogen, equivalent to 200 lb/acre, along with other complete fertilizers, prolong life.

One should be aware that some trees cause problems. Willows grow into sewer lines, their roots branching profusely into the nutrient-laden leakage until eventually they enter the crack in the pipe. Rapid growth then clogs the drain. Cottonwoods raise a 4-in. concrete slab in a short period of time—from a hairline crack to a 2-in. lift in a matter of weeks. Ginkgo, an Asian tree with fan-shaped leaves, produces a fruit in the fall that gives the species the colloquial name stink-bomb tree. The malodorous "prunes" of the female are, most politely, nauseous. Pine needles quickly fill roof gutters and release acids that hasten decay of wood-shingle roofs and encourage rusting of galvanized gutters. Trees that prune naturally, such as conifers, may drop branches on telephone and power lines. Some trees emit droplets of oily substances that discolor painted surfaces of autos and houses. Others, like the sweetgum and sycamore, drop ball-like fruit that some people consider a nuisance in the yard or patio.

Trees several years old and raised from seeds in cans, usually available at commercial nurseries, are often root-bound, the roots circling round-and-round the interior of the container. Unless carefully straightened and extended in their normal radial direction, the roots continue to grow circularly when out-planted. Then, later, the tree may topple with the wind, exposing a root system appearing like a gigantic ball of twine. Roots must be cautiously straightened, lest they break and the stem and its foliage lose these essential channels for movement of the soil's moisture and nutrients.

Transplanting trees presents problems. Those transferred from the woodlot have low survival rates. The transplant shock of trees more than 2 in. in diameter at the ground line makes moving them rarely successful, as an inadequate amount of undamaged root is lifted with the tree. Death may be delayed a year or even two; but in that interval of time, a smaller tree when planted, even with bare roots, would be a couple of feet tall.

Here are some rules of thumb for transplanting trees:

1. The roots must never be permitted to become dry. Even 30 sec. in the windy air will desiccate them.
2. No roots are allowed to be exposed following planting.
3. One is not likely to plant too deep for survival, but to plant too shallow invites seedling mortality.

Once the trees are well-established, exhibiting vigorous growth, pruning may be desirable. Some folks prefer to forego limb removal, allowing branches to hug the ground and to hide the house. Other people prune up the bole each year until the tree's branches are beyond reach, enabling one to walk under all the trees. This also fully exposes the home to view, a recommendation of the sheriff's department. Still others prune some species (like pines) in some places (like close to the house) and leave other kinds, such as dogwood and arbor vitae, untrimmed.

Shears are used for pruning smaller branches and a saw for larger ones. Never use an ax. The cut should be flush with the bole. When using a saw, the worker begins below the branch, cutting upward for a third of the branch diameter. Then, when the cut is completed from the top of the limb downward, the weight of the heavy branch will not tear the bark from the side of the tree as it falls.

It pays to use a pruning paint containing a fungicide when trimming broadleaf trees, for spores of wood-rotting fungi are always in the air. These spores, coming to rest on fresh wounds, "germinate" into long hair-like mycelia which grow up and down the bole, decomposing the wood with their secretions as they travel. The same treatment is suggested where dead branches are removed from conifers.

For pines, the resin flow from the wound where living branches have been removed serves as an effective scab to keep out airborne fungi spores. No paint is necessary: it may be harmful if its use precludes the forming of the resinous scab.

As to fertilizing yard trees, a single significant suggestion: double the dose underneath the crowns of the trees for the complete fertilizer used for the rest of the lawn. That is a good general rule where the homeowner has had a fertility analysis made by a state laboratory for a recommended nutrient amendment for the grass. While tree roots extend well beyond the "drip line" of a tree's canopy, most of the feeding roots will be sufficiently supplied with the abundant nutrients under the crown. Response to added "plant food" is many times better if adequate water is regularly supplied for easy absorption of the chemicals.

A good general fertilizer prescription for first-year seedlings is to broadcast at the beginning of the growing season about half a pound of a well-balanced complete fertilizer (8-8-8, 10-10-10, or another approximate formulation) in a band 2 to 3 ft wide, beginning half a foot from the tree. About the first of June, and again in mid-July, one cup of a nitrogen fertilizer (ammonium nitrate, nitrate of soda, ammonium sulfate or urea) is similarly applied. The amount of fertilizer and the radius of the band are proportionately increased each year until 6-year-old stems receive 3 lb of the complete fertilizer over an area with a radius of 15 ft at the first application. About 1 lb of the nitrogen material is used for the later applications. Mature trees annually require about 5 lb of a nitrogen compound.

Two myths need to be buried: (1) feeder roots are deep in the soil and (2) they extend underground only as far out as the branches reach. Most of the small, almost microscopic feeder roots are just under the soil surface, intertwined with grass roots. Grass roots usually extend much deeper than do the feeding roots of trees.

As plant nutrients readily leach into the soil and are quickly absorbed by roots, the common and expensive practice of first boring holes is unwarranted. Fertilizers simply spread on the surface are carried down in water solutions. In addition to being hard work, placing fertilizer chemicals deep in the soil is wasteful, as the elements are soon carried below the reach of tree roots. This practice allows the fertilizer salts to kill upon contact the roots injured by the boring. It also places the fertilizer nearer to the groundwater, which it contaminates, and is lost to the tree. The procedure is expensive. Simply broadcast and water-in well.

Cultivating plant food into the surface soil, like hole-boring, is needless effort. It is also undesirable because many of the fine rootlets which absorb the elements are severed and destroyed. Entrance of dry air further inhibits root growth and nutrient absorption. Because roots extend well beyond the crowns of trees—often exceeding 50 ft for young pole-size trees—fertilizers could be spread beyond the span of a tree crown.

No suggestion is made here for liming, for some species—like the pines—do best in acidic soils (pH 5.5), while others—like the junipers—prefer soils approaching, or on the alkaline side of, neutral (pH 7).

Johnny Appleseed's scattering of spiritual seeds and apple seeds along the trails he traveled cheered the pioneers on their toilsome way. So, too, trees in the landowner's yard—urban and suburban—will rouse spirits and provide cheer for those who pass by.■

STEWARDSHIP

Altruistic care of the forest based upon an understanding of ecology and economics.

C H A P T E R 4 5

SOME BASIC ECOLOGY

Much has been made of the derivation of the term "ecology" from the Greek word *oikos*, meaning house. Thus the ecologist is concerned with his habitat and its care. In contrast, little has been made of the translation of *oikos* into Latin and the word derived therefrom. It is *i(e)conaea*. Of course, the English "economics" is its offspring. Little, too, is made of the translation, by King James' Bible scholars, of both *oikos* and *iconaea* into Anglo-Saxon as "steward" and "stewardship."

Ecology and economics are inseparable in stewardship. Principles of both disciplines concern the ecologist as the manager of the household or the keeper of the accounts. As a faithful steward, the landowner acts accountably in caring for the resources entrusted to his supervision. As silviculture relies upon ecological principles, its practice, too, becomes closely allied to stewardship.

Perhaps for no other natural resource is the relevance of stewardship more appropriate than for forests. Most landowners take seriously the Genesis directive of Hebrew-Christian persuasion to "replenish and subdue the earth." To subdue is to take charge, but to replenish implies the restoration of exploited sites as well as the care of lands used to provide for the needs of men.

Ecology, defined as the study of the interrelationships of living things to each other and to their environment, has significance in forestry. Important timber species normally occur in ecological associations cataloged as forest cover types. Some species are important in several types. While certain ones grow successfully when introduced outside of their native range, the use of exotics may be ecologically disruptive.

Large trees are old by human standards. One should not, however, assume that they are necessarily permanent or invulnerable to death and decay. All timber stands reflect stages in an ordered sequence of ecologic succession which may be interrupted anytime by natural catastrophes, such as windstorm, insect epidemic and fire, or by logging. To produce the stable, or climax, forest, ecologic succession often requires centuries without natural or human disturbances.

Most virgin forests of lumbering history were subclimax, or at even earlier stages, when harvests for building materials began. For example, white and red pines of the North, the yellow pines of the South, and the Douglas-fir of the West seed in abundantly only on denuded sites. Yet these species were ready for harvest in the United States in the 18th, 19th, and 20th centuries because of earlier interruptions in ecologic succession. Indians burned

Jack pine, a serotinous species, and aspen initiated new forests following the cut-out-and-get-out harvests and the extensive wildfires that occurred throughout the Lake States until about 1900, *U.S. Forest Service photo.*

the forest and lightning fires and hurricanes laid bare the land. In time, other species, perhaps less valuable economically, would have replaced them, as we shall see.

Even the southern baldcypress of the swamp may have had its start following fire. Oaks and hickories, not dependent upon fire to scarify the soil, enter the woodland scene five or 10 years after a pine forest is established.

The key to forest succession is the relative tolerance of trees to competing demands by their neighbors for light, moisture, and nutrients. Species like sugar maple, beech, and the hemlocks persist under dense shade. Regenerating under their own canopies, these kinds dominate the climax forest.

At the other extreme are species intolerant of shade, including most pines and many hardwoods. Aspen, cottonwood, willow, and yellow-poplar, even though prolific seed-producers, compete poorly with established vegetation. They are rarely found in understories. Their seeds germinate readily on exposed mineral soil and their seedlings grow rapidly, if light is abundant, to "capture" growing space and exclude competitors. Stands of these species usually result from logging, abandonment of cultivated fields or where new river silt has been deposited.

Seeds of most plants appearing in ecologic succession are usually available when needed to fulfill their tasks. Thus, a disturbed site may be dominated the first season by annual weeds or grasses, followed successively by biennial and perennial herbs and shrubs. Eventually trees encroach. On fertile sites, some of the initial annual invaders may grow so thick as to inhibit establishment of the pioneer tree species, delaying the development of a tree cover until a new catastrophe exposes the site to full sunlight.

Stands of tolerant species, as noted earlier, are usually climax types, represented by beech, sugar maple, spruce, and hemlock in the Northeast, oaks and hickories in the Central States and the South, and redcedar, spruces, and firs of the Northwest. Harvesting individual trees in these forests by the selection system provides space for new stems and perpetuation of the type. While the single-tree selection system is often effective in complex types such as those of the southern river bottomlands, encroachment of weed species in the openings frequently precludes using the method.

An extensive forest cover type in the North consists mainly of sugar maple, American beech, and yellow birch. Its simplistic composition contrasts sharply with the multiplicity of species—perhaps over 50—in the mixed mesophytic forests of the South.

This maple-beech-birch type is usually climax because all three species are tolerant of shade. On moist sites, however, the more tolerant eastern hemlock may invade and become dominant. Birch, although the preferred species commercially, is less tolerant than its associates. Its light seeds, widely distributed by wind, and its poor deer browse encourage seedling establishment. Beech, least desirable because of fungus infections that cause rot, is most tolerant and has heavy seeds distributed mainly by animals. Maple is intermediate to its two associates in both tolerance and seed dispersal.

On most sites, the maple-beech-birch type can be perpetuated by selection harvests. This management system, recommended where stands include trees of many ages, causes minimal disturbance to soil, watersheds and esthetic values, while providing small openings for wildlife browse. It favors the more tolerant beech, affording a greater number of hollow (because of rot) den trees and beech nuts, a key wildlife food. A chief drawback is nature's tendency to favor the less-useful beech over the preferred birch and maple. It is also costly to log in this manner. Trees are often damaged, thus encouraging disease infection. Selection, however, affords opportunities to correct undesired ecologic trends by harvesting less-desirable stems at frequent intervals.

White pine, long the most valuable timber of northeastern America, was found as large, high-quality trees in many of the virgin forests of New England, the Lake States and the Appalachians. It is not, however, a climax species. Being less tolerant than most of its associates, white pine is unable to regenerate under its shade. This pioneer tree seeds-in on burned-over, cutover, or abandoned farm lands. Long-lived, individual stems have stood in undisturbed situations until ecologic succession has surrounded them with climax species such as eastern hemlock.

Because of its intermediate tolerance, white pine can be regenerated by silvicultural systems ranging from individual tree selection to clearcutting. The shelterwood system—removing the old forests in several harvests—and clearcutting—simulating nature's hurricanes and lightning-caused fires—remove enough tree cover and sufficiently scarify the soil surface for pine seeds to germinate and seedlings to thrive.

Regeneration of white pine by selection, favored by some because it is least disturbing to soil, watersheds and esthetic value, is feasible primarily where associated species bear sparse foliage and are generally less tolerant than these pines. Such intolerant associates include paper birch, red maple, jack pine, aspen and pin cherry. Costly manual or chemical release from this competition is often essential, making selection cutting uneconomic. Single-tree selection also tends toward site domination by species with more tolerance but less economic utility than white pine.

Shelterwood cuttings are designed to achieve regeneration gradually, over periods of 10 to 30 years. A first cut stimulates seed production, allowing seeds to fall on scarified soil. This enable seed germination. About 60% of the trees are left standing. A decade or so later, a second harvest removes 60% of the remaining trees, releasing the seedlings that germinated after the initial harvest. Part of the parent stand continues to be left for shade to protect the seedlings from sunscald. A final cut, after regeneration is established, removes the rest of the old stand.

The many species of southern pines produce essentially identical wood, marketed as a single material. The principal ones occur in pure stands, in mixtures with hardwoods, and with each other. None is a true climax species, for on most sites their seedlings are unable to survive under the shade of their parents or that of invading competitors.

Four species are of primary importance. Loblolly pine, the most widespread and abundant, is a fast-growing tree of intermediate tolerance highly regarded by timber growers. It now dominates much land formerly occupied by longleaf pine. Shortleaf pine extends farthest north, is intolerant, and outproduces its hardwood associates on many sites. Slash pine, fast-growing and on some sites relatively tolerant of competition, has been widely planted; genetically superior strains are available. Longleaf pine is most intolerant, grows slowly in youth, but develops straight, clean stems of excellent quality.

In an earlier day, longleaf pine timbers went to the shipyards of Greece for masts. Longleaf and slash pines occurring mainly on the Gulf and Atlantic coastal plains produce commercial oleoresin as well as fiber.

The ecologic climax for most southern pine sites is a hardwood type, with occasional old-growth relict pines. Such forests occurred widely in the Piedmont Province in pioneer days. Many dry sandy sites now growing pine would, without disturbance, succeed to a forest of scrubby hardwoods. Windstorms and forest fires, under natural conditions, interrupted the ecological cycle to initiate the extensive pine stands of the South. Despite wasteful cutting and woods-burning, present-day pine stands almost always originated as natural reseeding of cutover forest or abandoned farmland.

Management of southern pines requires maintenance of a sub-climax vegetative association. The alternative, allowing stands to revert to predominantly hardwood types, is incompatible with optimum production of structural lumber, plywood and long-fiber pulp.

Managed southern pine is currently harvested under systems that remove most or all of the existing stand, permitting control of established hardwoods by fire, mechanical means, or both. Harvests may leave five to 10 seed trees/acre, be limited to strips or patches that will reseed from adjacent timber, or depend on artificial planting or seeding.

Douglas-fir, one of the major timber species of the world, reaches its best development in the virgin forests west of the Cascade and Sierra Nevada ranges from northern California to southern British Columbia. Most current harvesting and reserves of the species are concentrated in this region. The Douglas-fir found in Rocky Mountain forests is a distinct and generally less-productive race of the species.

Compared to other western conifers, Douglas-fir is intermediate in tolerance, but in comparison with its associates in the same forest, it is intolerant. It cannot regenerate under the shade of the more tolerant redwood, western redcedar, hemlock, spruces, true firs and hardwoods with which it competes. Ecologically a fire-subclimax species, Douglas-fir owes its dominance to fire history, windthrow and to its longevity. Disturbance provides bare soil for seed germination and openings where seedlings can outgrow competitors. A head start and fast growth enable its crowns to retain dominance for 350 years.

Faster growing and intrinsically more valuable than its associates, Douglas-fir best utilizes the land when grown in pure stands. Clearing of some kind is essential if regeneration is to be accomplished by natural means. Partial cuttings also expose residual stems to wind, precluding shelterwood and seed-tree harvests in dense virgin and second-growth stands. (Other factors that make clearcutting almost unavoidable are the heights and weights of the trees, necessitating large logging equipment which is difficult to maneuver among uncut stems without injuring them.)

Most foresters of the Douglas-fir region consider clearcutting in relatively small patches the best approach to regeneration of existing old-growth stands. Prompt burning of logging debris after harvest reduces wildfire hazard and enhances seed germination. Small blocks permit adequate seed production and dispersal from adjacent stands, a marked contrast with earlier extensive logging practices where the dearth of seed precluded adequate regeneration. Small blocks also reduce adverse hydrologic effects by limiting the proportion of a watershed which is disturbed.

As the Douglas-fir region shifts from virgin to second-growth and managed forests, changes in harvesting methods are likely to occur. Trees will be cut at younger ages—50 to 75 years *vs* 150 to 350. To promote high-value growth, fewer stems will be grown per acre. As these trees will be smaller and more wind-firm than those in virgin stands now being harvested, partial cutting systems may be feasible, especially where wildlife, watershed and scenic values are paramount. For most commercial sites, however, the trend seems to be toward small block clearcutting followed by planting or direct seeding to insure prompt and adequate restocking.■

C H A P T E R 4 6

ETHICS OF FOREST USE

Carl Schenck, founder of the Biltmore Forest School in 1898, the first such institution in the New World, expressed to his students a concern for altruism with the phrase, "Excelsior, the higher good." That sums up the matter of environmental ethics about which this chapter addresses the landowner of small forested tracts.

In the conquest of the land of the New World, the Puritan refugee settlers found nature obstinately defiant of efforts to tame it. "The woods and thickets," the governor at Plymouth wrote, "represented a wild and savage hew." They still do to those who challenge their thorns, seed ticks, copperheads and widow-makers. The forest is to be tamed and used. Its taming was described two centuries later by Walt Whitman in the *Song of the Broad Axe.* The axe leaps, and from the forest come shingle, rail, sash and floor. And more.

Materials used by people mandated utilization and consequently the management—or taming—of these lands. People use paper for communicating, packaging and sanitation; lumber for housing and bridging; poles for utility lines; pilings for foundations for society's building structures; fuel for furnaces; and raw materials for chemicals. A century ago, when Whitman described the woods, civilized peoples depended almost solely upon wood for housing (roofs, window frames, siding), for fuel (locomotives and steamships as well as hearths), structures (mills and mines, as well as offices), trains (tracks as well as water towers, crossties and trestles), woodenware (table plates and eating utensils, buckets and barrels) and chemicals (potash for soap, alcohol, turpentine and rosin).

In time, partly because of the cut-out-and-get-out experiences of the timber barons in the Northeast, these woods would be romanticized. The Transcendentalism of the period, the worship of God in nature, is with us yet. So, too, is the obstinately defiant forest from which must come the sash and floor. This is the philosophical dichotomy—the paradox—with which the landowner must come to grips. Altruism, that uncalculated devotion to the interest of others usually accorded to an ethical principle, is both preserving nature and providing material for man's comfort. Romanticism and utilitarianism must yet be wed by wise resource management, or stewardship. That kind of taming of a natural resource is altruistically motivated.

In the time of Henry David Thoreau and his associates, we assume from their writings, even private lands were open for all to enjoy. Today's wilderness enthusiasts see this as a dogma in an unwritten code of environmental ethics. The more radical

This white pine tree stood long following the cut-out-and-get-out days that ended in New England around 1860 as the migratory lumbermen moved west to the Lake States. The famous landmark tree now overshadows its progeny, *U.S. Forest Service photo.*

among them lie down in front of herbicide-dispersing tankers on private lands, walk through a hunter's lease at dawn during deer season beating on cans and sue the federal government to prevent the owner of timber that stands on national forest lands from harvesting his purchase.

Other environmental problems faced by resource managers, like soil erosion and depleted range, involve the question of short-term versus long-term economics. Present management practice which is less than the best may add greatly to the costs of food and fiber in the future. Good practices now, however, increase the price of commodities for today's consumers.

The following examples provide further illustrations of the complexities of environmental policies. Congressional action makes permissible the regeneration of trees on national forest lands by the most expeditious scientifically accepted methods. For many commercially valuable species, this means clearcutting. This practice produces landscapes which are ugly for a few years, but other methods double the cost of logging, and that cost is passed to the consumer.

Nature may be "awful-ly" ugly as well as "awe-full-ly" beautiful. Black widow spiders, cancer and (to some) the slowly decaying timbers lying under a high-country stand of western redcedar are not things of beauty. A compulsive "neatnik," for instance, could not tolerate the accumulation of organic debris that awaits fungal action. Thus the dichotomy of Man's attitudes toward Nature—a thing of beauty and yet sought for its utility—is apparent.

Placing responsible utilitarianism above romantic sentiment requires Man to accept the responsibility, not just the authority, to harvest ripe timber in order to satisfy the needs of people. It is in fact ethical to do so. It is the meaning of dominion, even dominion over a delicate creation.

To maintain the delicate balance of creation calls for sacrifice, rather than self-service. Most religions remind one of the need to reform lifestyles to demonstrate harmony with the earth and a stewardship that stresses the major elements of environmental conservation. But we must be careful to avoid "Man's power over Nature turn(ing) out to be a power exercised by some men over other men with Nature as its instrument," as C.S. Lewis pointed out.

A recent assignment in Indonesia vividly demonstrated this principle to the author. Each governmental decree issued in an effort to prevent exploitation of resources by foreign interests inevitably backfired to economically harm the very citizens—those on the lowest rung of the social ladder—it was intended to benefit. Foreign industries, rather than subscribe to the directives, went home. Their exodus left peasants without employment. The land, to be sure, will look more appealing in its natural, virgin state. But what about the people? They hunger for employment that consumes resources.

Thus, the real "tragedy of the commons" is Man's egocentricity which is hardly to be trusted for environmental care. If laws and taxes cannot enforce conservation, martial repression and government-imposed rationing, in contrast to the marketplace, may be expected to handle the task.

The need for ethical reminders like this chapter is because of greed. Theology challenges us to cast aside our selfishness in the use and abuse of resources in order to see clearly our weaknesses as stewards. Will we be *Homo "egoiens"* or *Homo sapiens?* As the latter, stewards of natural resources take a worldwide view of the biblical injunction to love their "neighbors," those in every age and clime for whom the woods and wildlife are now cultured and husbanded.■

AFTERWORD

For some lands on this country, human history has altered the course of ecologic transition. Never, perhaps, has the use of land been altered so rapidly by man in so short a period as within the past century.

Man's use of the land and its resources often sets in motion natural processes that cause permanent change. Frequently the results are destructive; sometimes they are catastrophic. Erosion from cleared slopes removes topsoil developed over thousands of years. Gullies from poorly engineered roads lower the moisture regime of adjacent soils and degrade forest sites. Although the forests of the nation have shown remarkable ability to reoccupy lands clearcut for timber, burned or cleared for agriculture, new stands are usually of different species composition than were those established under virgin conditions. Pines, sweetgum, yellow-poplar, or scrub oaks dominate rehabilitated sites where once-vigorous stands of beech, birch, maple, American chestnut, or the oak-hickory forest types covered the land. Hence, the forests in much of the contiguous 48 states have been modified more or less severely by previous land use.

Forestry practices, as well as the vegetative composition of the forest, will continue to be affected by land-use transition. Conversion of timber-producing lands to residential communities, factories and reservoirs; or withdrawal of forests for watersheds, game preserves, parks and research areas will necessitate intensifying practices on remaining lands to supply the nation's fiber needs. The need for highest production on the land will rarely permit reversion to original timber types. These managed forests, however, will be subjected to the same geographic and related climatic influences as their predecessors. Economic influences, too, may be little different.

The landowners who manage the woods will be most successful when they are guided by wise consideration of the best management practices for assuring clear, clean water and other amenities, as well as for wood fiber. Hopefully, even Adam Smith, the 18th-century Scotsman who penned the *Wealth of Nations* which gave the populace an understanding of capitalism, would subscribe to them.■

GLOSSARY

These definitions are taken from the author's *Trees: An Introduction to Trees and Forest Ecology for the Amateur Naturalist*, published by Prentice-Hall, Inc.

abscission Layer of cells that, in growing, separate a leaf petiole from its twig.

acidic soil A high hydrogen-ion concentration in the soil solution, acid reaction, below pH 7.0, referred to as "sour."

acorn Usually one-seeded fruit of oak trees, with a hard wall and generally partially enclosed in a husk.

adventitious bud Buds arising at positions, other than where leaves or stems ordinarily arise, such as on roots, at the base of trees, and often as a response to wounding.

airplant Epiphyte; a plant growing upon another of a different kind, though not parasitic, deriving nutrients and moisture from the air.

alkaline soil Soil high in hydroxyl (OH) ions, associated with high calcium levels, pH above 7, hence basic or "sweet."

alkaloid Organic substance with alkaline properties occurring in plants.

allelopath In plant ecology, an influence, usually chemical, of one plant (other than microorganisms) upon the growth and vigor of another.

alluvial Soils developed from water-transported material and accumulated in delta-like fans or on lands of river overflow.

anaerobe Bacteria and other organisms that live in the absence of oxygen.

angiosperm Plants with seeds in closed ovaries, such as broadleaf trees.

anion Negatively charged nutrient element particle.

annosus root rot A fungal rot occurring chiefly in the roots of coniferous trees.

arboretum Place for cultivating trees and shrubs for scientific or educational purposes.

auxin Natural growth hormone in plants.

backcross Procedure for breeding a hybrid with one of its parents or parental types.

bacteria One-celled microorganisms active in fixing atmospheric nitrogen, producing disease.

basal area/acre Total area, expressed in square feet, of the cross sections at breast height of all the trees on an acre.

batture Bottomlands occupying areas between rivers and levees and usually flooded yearly.

biome An ecological community of interdependent organisms convenient to recognize and describe as a unit.

birdseye Figure produced in wood by small conical depression of fibers, appearing as little eyes.

board foot Unit of measure represented by a board one foot long, one foot wide, and one inch thick.

bog Uncultivated tract characterized by poor drainage, acidic peat, and low vegetation.

book-matched In veneer, sheets matched by turning over alternate slices of wood.

buckshot Clay soil structure appearing like large cubes or like the lead shot of shotgun shells.

bull-of-the-woods Supervisor of a logging camp.

Bundle scar Mark left in a leaf scar by the severing of vascular bundles (within the petiole) at leaf-fall.

burl Hard, woody excrescence on a tree, usually resulting from the entwined growth of a cluster of dormant buds.

burr Colloquial for pine cone; sometimes bur.

cache A store of items for safekeeping, such as where squirrels hide acorns.

cambium Living tissues between xylem (wood) and phloem (inner-bark) of woody plants.

canopy All the green leaves and branches formed by the crowns of trees in a forest.

cation Positively charged nutrient element particles.

catkin Flexible, scaly spike bearing flowers of one sex.

cellulose Complex carbohydrate compounds occurring in wood and other plant material.

chaparral Thicket of shrubby evergreen oaks; now includes non-oak species.

check In wood seasoning, a longitudinal fissure caused by fibers separating along the grain.

chlorosis Yellowing of foliage, symptomatic of a nutritional deficiency.

climax Final stage of ecological succession: the plant community that continues to occupy an area as long as climatic or soil conditions remain unchanged.

clone A group of plants derived by asexual reproduction from a single parent.

cohort In ecology, all individuals that function in synchrony, such as when all plants of the same inheritance flower simultaneously.

colloidal Matter of small size (less than 2 microns) and having high surface areas per unit of weight.

competition Whenever several organisms require the same things in the same environment; "rivalry" between plant species for control of a site.

compression wood Abnormal wood formed on the lower sides of branches and inclined trunks of coniferous trees.

cone Fruit with overlapping scales.

conelet Immature cone.

conk Visible fruiting body of wood- or tree-destroying fungus, often projecting from a tree trunk.

controlled fire Any deliberate use of fire on land whereby burning is restricted to a predetermined area and intensity.

cooperage Wood for barrels, consisting of two round heads and a body composed of staves held together with hoops.

coppice Method of renewing forest in which reproduction is by sprouting.

cotyledon First leaf arising from a seed; seed-leaf.

cover type Pioneer, temporary, or climax combinations of forest trees.

cupule A cup-shaped sheath characteristic of the partial cover of an oak acorn.

cutting Segment of stem cut from trees and used for vegetative propagation.

dbh See *diameter, breast height.*

dendrochronology The study of growth rings in trees to determine and date past events.

dendrology Identification and systematic classification of trees.

diameter, breast height (dbh) Diameter of tree 4.5 feet above average ground level, used for volume and growth determinations.

dicotyledon An angiosperm in which seedlings have two seed-leaves.

dioecious Trees in which male and female flowers are produced on different plants.

dominance Pertaining to the species most characteristic of a habitat which may determine the presence and type of other species.

dominant Trees with crowns extending above the general level of the crown cover and receiving full light from above and at least partly from the side.

dormancy Resting stage of a plant, when a tissue predisposed to proliferate or develop does not do so.

drupe Simple one-seeded fleshy fruit with bony inner-wall.

duff Forest litter and organic debris in various stages of decomposition.

ecology The study of the interrelationships of living organisms to each other and to their environment.

edaphic Referring to the soil.

eluvial Pertaining to a soil horizon from which minerals, often colloidal in size, leach to a lower enriched zone.

endangered Class of plants or animals in immediate jeopardy of extinction; more serious than threatened.

escape An introduced exotic plant later found growing wild.

estuary Frequently flooded land where a river meets the sea.

even-aged Applied to a stand of timber in which relatively small age differences exist between individual trees.

face, naval stores See *face, turpentining.*

face, turpentining The exposed, debarked surface of the tree from which oleoresin is collected. A tree may have one or two faces.

faller One who fells timber.

"fat" pine Conifer wood, usually heartwood of stumps and logs of the virgin forest, abnormally flammable because of high resin content.

fiber length The long dimension of fibers of which wood is composed.

fibro-vascular bundle Specialized conducting tissues in wood.

financial maturity Age at which it is advisable economically to harvest trees, individually or as a stand, in contrast to physiological maturity.

fire climax Species that will continue to be maintained on an area if fires occur at appropriate times.

fire, crown A fire that runs through the tops of trees.

fire, ground Where organic matter in the soil is consumed, especially in peaty soils; fire may burn for long periods entirely below the surface of the ground.

fire, surface A fire that burns only surface litter and small vegetation.

forb Non-grasslike herbaceous plant; herb or weed in stockman's language.

forester A person professionally educated in the management and wise use of renewable natural resources, including timber, water, range, recreation, and wildlife.

frontland slough Shallow depressions adjacent to the banks of former stream courses in which water collects but normally disappears during the first half of the growing season.

frost-heaving Upward displacement of seedlings due to expansion of ice in the soil.

fruit Seed-bearing product of a plant.

fungus Plant, serving principally for decomposition, without chlorophyll, roots, stems, or leaves. Some do not play a decomposing role: mycorrhizal growths on roots and the nongreen component of lichens on rocks are examples.

gall Tumor or pronounced growth of modified tissue, caused by irritation by a foreign organism such as an insect.

gallic acid Organic compound in plants, especially in galls and teas.

genotype Resulting organism of the sum of all the hereditary genes in an individual.

geotropic Characteristic of a plant root to grow downward.

germination Rupture of seedcoat and concurrent development of rootlet (radicle) and leaves (hypocotyl).

glacial, glaciated Pertaining to the action of ice upon the land during the Ice Age.

glacial outwash Usually sand plains formed from water-transported material during the ice age.

glade Grassy, open space in a forest.

gumbo A fine, silty soil, common in the southern and western U.S. that forms an unusually sticky mud when wet.

gymnosperm Plant bearing naked seeds.

heartrot Decay generally confined to the heartwood of living trees.

heartwood Inner core of a woody stem, composed primarily of nonliving cells and usually distinguished from the outer enveloping sapwood layer by its darker color.

heliophyte A plant thriving in, or tolerating, full sunlight.

herb Non-woody, flowering plant, including broadleaf forbs and grasses.

herbicide Chemical for killing plants; in forestry, sometimes dendrocide or silvicide.

high-grading Removal from a forest of only the highest-quality trees, leaving lesser-quality stems for future harvests and as a source of seed.

high wheels Large wheels used for transporting logs by raising them off of the ground, from felling site to loading site, pulled by animals.

host (biological) Organism upon which another organism feeds and develops.

hummock (hammock) Low area with deep rich soil.

humus Plant and animal residues that are undergoing decomposition in and on the surface of the soil.

hybrid Cross usually between two species.

hydric For soils and sites, meaning wet conditions prevail.

hypsometer Instrument for measuring height.

illuvial Pertaining to a soil horizon into which minerals, often colloidal in size, have been leached from an impoverished zone.

increment borer Auger-like instrument used to extract radial cylinders of wood that show annual growth rings.

indicator See *plant indicator.*

inflorescence Character of floral arrangement.

Inland Empire Region between Cascade range and Rocky Mountains in Washington, Oregon, Montana, and Idaho.

internode Portion of the stem between two nodes; the clear trunk between the zones where branches develop.

lamma Shoot formed after a pause in growth; shoot formed on a tree bole when suddenly exposed to light.

larva The wingless, often wormlike, form of a newly hatched insect before undergoing metamorphosis.

lateral root Roots extending horizontally from the taproot or the base of the tree.

leaf scar Mark left on a twig at the point from which a leaf falls.

legume Dry fruit; product of plants of family *Leguminosae.*

lenticel Cells loosely arranged on the outer layer of twigs and serving for the exchange of gases through the otherwise impermeable surface.

lesion A circumscribed disease area on a trunk or stem.

levee Embankment, natural or manmade, to prevent flooding.

lignin Second most abundant constituent of wood, a part of the thin cementing layer in woody cell walls.

loam A mixture of sand, silt, and clay in proportions optimum for plant growth.

low flat Terrain between ridges, usually 2 to 15 feet lower in elevation, in river bottoms.

lumber Product of a sawmill or planing mill, and not further manufactured.

mast Fruit of trees considered useful for food for livestock and wildlife.

mature An individual tree or stand for which full development has been attained; may be physiological, economic, or sexual.

meristem Tissue capable of cell division and, therefore, of growth.

mesic For soils and sites, meaning moist but well-drained.

mineral seedbed Exposed soil exclusive of organic matter.

monocotyledon An angiosperm in which seedlings have a single seed-leaf.

monoecious Plant with male and female organs in different flowers on the same plant.

mor Soil whose upper mineral layer is relatively free of organic matter but above which is organic matter in various stages of decay.

moraine Deposit of glacial drift of rock and small sediment at the base or sides of a glacier.

muck Fairly well-decomposed organic soil material, containing mineral matter and accumulated under conditions of imperfect drainage.

mulch Material such as straw, leaves, sawdust, or paper, spread on a soil surface to retard water loss and weed growth.

mull Soil whose upper mineral layer is intimately mixed with organic matter.

muskeg A moss or peat bog, usually partly forested.

mycelium Collective term for minute hyphae strands or hairlike filaments of a fungus.

mycology The branch of botany that deals with fungi.

naval stores (gum) Historically and currently products of the resin industry, particularly turpentine and rosin.

nematode Minute wormlike organisms attacking roots of plants.

nitrogen-15 Form of nitrogen that can be traced in vascular systems using radiation technology.

node (stem) Point on a stem that bears a leaf or leaves.

nodule (bacteria) A growth, as on the roots of legumes and certain other plants.

oleoresin Natural resinous substances and oils occurring in or exuding from plants.

osiers Willows having long, rodlike twigs used in basketry and fencing.

overburden Material overlying a former surface soil or a mineral seam.

pales weevil A beetle of the genus *Hylobius* that attacks the inner-bark of certain pines, especially seedlings, immediately following logging, as it breeds in freshly cut stumps.

pallet Wood platform on which material is stacked for convenient handling and transport.

particulate Minute solid and liquid pollution matter in the atmosphere.

pathogen Organism capable of causing disease.

peat Soil consisting largely of undecomposed or slightly decomposed organic matter, accumulated under conditions of excessive moisture, and containing little, if any, inorganic matter.

peduncle Stalk of a flower.

permeability, soil Relative rate of penetration by a solid object or by water, influenced by particle size, structure, and moisture content.

petiole Stalk of a leaf supporting the expanded portion or blade.

pH Logarithm of the reciprocal of the hydrogen-ion concentration, indicative of soil acidity or alkalinity (7 is neutral).

phenology Science dealing with the time and appearance of characteristic periodic phenomenon in a life-cycle of an organism, as in flowering of plants.

phenotype External characteristics of an organism due to the interaction of its genetic constitution and the environment.

phloem Inner-bark: principal tissue concerned with movement of carbohydrates in plants.

photosynthesis The manufacture of carbohydrate food from carbon dioxide and water in the presence of chlorophyll.

phototropism The tendency of a plant to grow toward the light, as a tree grows skyward.

phyllotaxy Arrangement of leaves on a twig.

physiology Deals with the life processes of organs, within living organisms, as the root of a tree.

piling Sound timber driven into the ground to support structures (buildings, piers, and so forth).

pioneer Those plants that originate successional patterns, as lichens and mosses on rock surfaces, or certain shrubs or trees following fire.

pistillate Pertaining to the female flower or inflorescence.

pitch pocket A concentration of resin, derived from sap, in wood of conifer trees.

pith Primary tissue in the central part of a stem, twig, or root.

plant indicator Plants that, by their presence, denote site productivity for forest trees.

plow sole the depth to which the plow continuously turns the soil.

plywood Three or more layers of wood veneer joined, with grain of adjoining plies at right angles, with glue.

pocket gopher Short tailed, burrowing mammals of the family *Geomyidae*.

pocosin Swamp in a slightly elevated area in the southeastern Atlantic states.

podzol Soil characterized by leaching of iron and aluminum oxides from a surface horizon to one below, leaving an ash-gray stratum near the surface.

pole (product) Round timber suitable for supporting utility lines or crude buildings.

pole (silviculture) Tree of median maturity and size (4–12 inches d.b.h.).

pollard Shoot produced when a tree crown is systematically cut back beyond the reach of browsing animals and which has commercial value.

pollen Fine, powdery grains, also called spores, produced in male flower parts and disseminated, often by wind, to female flower parts where the sperm develops following germination.

pollination Transfer of pollen to receptive part of female flower.

polycotyledon A gymnosperm in which seedlings have many seed-leaves.

polytrichum moss Genus of mosses, chiefly of temperate and arctic regions.

popples Colloquial name for poplar trees.

pores, soil Space in the soil between solid particles,

filled with water and/or air.

porous wood That which in cross section exhibits vessel or vascular elements, typical of dicotyledonous trees.

prescribed fire A controlled fire ignited under rigidly specified conditions of weather, soil moisture, time of day, and so forth, so as to result in the heat and spread required to accomplish specific silvicultural objectives.

Public Domain, The Land belonging to the government, formerly available for homesteading and other uses.

pulpwood Trees or wood suitable for manufacture into wood pulp.

pumice A porus, lightweight volcanic rock.

puncher Ox-, mule-, or horse-driver in a logging operation.

pustule Blister on the bark of a tree caused by fungal disease.

radicle Root of the seed embryo, from which develops the main root of a tree.

resin Organic substance bled from trees, usually pines, for gum naval stores.

root nodule Swelling produced on root of leguminous and certain other plants in which bacteria convert nitrogen from the air into forms usable by plants.

root sprout Shoot arising from adventitious bud on exposed or buried root.

rosin Hard resin left after distilling volatile oil of turpentine.

rough Accumulation of living and dead ground and understory vegetation, especially grasses and leaf litter, sometimes with underbrush such as palmetto or gallberry.

sapling Young tree beyond the seedling stage, usually 2–4 inches dbh.

sapwood Exterior wood, usually of pale color, in contrast to the more central heartwood of a tree.

sawpit A hole in which the sawyer stood and worked in the early-day mills, as the log was drawn across the hole's opening.

sawtimber Trees suitable for lumber or plywood.

schist Medium- to coarse-grained metamorphic rock composed chiefly of quartz and mica.

scion Unrooted part of a plant used for grafting to a rootstock.

second growth Trees that cover an area after the removal of the virgin forest, as by cutting or fire.

seedbed The soil or forest floor on which seeds fall.

seedcoat Hard covering that encloses the embryonic structure of a plant.

seedling Tree between seed germination and sapling stages; the product of a forest nursery prepared for out-planting in a field or forest.

seed trap Device used to catch seeds for analysis of abundance, and so forth.

seed tree Tree retained in silvicultural harvests to provide seeds for reproduction.

selfing In genetics, pollen from a plant disseminated to the pistil of the same plant.

serotinous Cones that remain closed on the tree unless subjected to high temperatures.

shade-tolerance Capacity of a tree to develop and grow in the shade of other trees.

shake Section split from bolt of wood and used for roofing or siding.

shelterbelt A strip of living trees and/or shrubs that provides protection from wind, sun, and snowdrift for open fields. See *windbreak*.

shelterwood harvest Removal of mature timber in several cuttings, extending over a period of years equal usually to not more than one-quarter of the time required to grow the crop, by means of which the establishment of natural reproduction under the partial shelter of seed-producing parents is encouraged, resulting in an even-aged stand.

shingle Thin, oblong piece of wood with one end thinner than the other, for covering roofs and outer walls, usually sawed.

shiplap In lumber, a board with reduced thickness along the edge to receive the edge or tongue of another piece.

silviculture Art of growing trees in managed stands for the production of goods and services; concerned with forest establishment, composition, and growth.

site A particular area, denoted by specific biotic, climatic, physiographic, and edaphic (soil) factors; a reference to land productivity.

site index Expression of forest site quality based on the average total height of the dominant and codominant trees in a stand at 50 years of age.

site-preparation Removal of unwanted vegetation, stumps, and logging slash prior to reforestation.

spikes (flower) Floral arrangement consisting of a central axis bearing a number of sessile flowers.

spore Simple, one-celled reproductive structure in certain plants, as the fungi.

sporophore Spore-producing organ, usually the larger fungi or pollen.

stagnation Condition of a stand of timber that ceases to make height or diameter growth.

staminate Pertaining to the male flower or inflorescence.

standard Tree of good growth and form, usually 12–30 inches d.b.h.

steam donkey Mechanical rig for transporting logs from place of harvest to position for loading on rail cards, trucks, or wagons.

stratification Method of treating seeds, often by placing them between layers of moist sand or peat, to overcome dormancy.

strobile Cone with overlapping scales.

stump sprout Shoot from dormant buds at the base of a tree.

succession Progressive development of vegetation toward its highest ecological expression, the climax.

sucker Shoot from lower portion of stem or from a root.

sunscald Localized injury to bark and cambium caused by sudden increase in exposure of stem to high temperature and, perhaps, sunlight.

super-tree Phenotype believed to be especially superior in inheritance to a particular characteristic, such as growth or form; also plus-tree, select-tree, or superior-tree.

suppressed Trees with crowns entirely below the general level of the crown cover, receiving no direct light either from above or from the sides.

sustained yield Continuous production, achieving, at the earliest practicable time, an approximate balance between net growth and harvest.

synchrony In ecology, the simultaneous occurrence of an event by all plants of the same inheritance, regardless of location.

taiga Subarctic coniferous forests of small trees that form a transition zone between dense forests and tundra.

tannin Water-soluble compound in certain plants, used in leather tanning and dyeing.

tap root A large primary root, extending downward, usually providing support.

temporary species Plant that gives way to other species in ecological succession of plant communities.

texture, soil The relative proportion of the various size groups of individual soil particles, as sand, silt, or clay.

tip moth An insect of the genus *Rhyacionia* whose larvae hatch from eggs deposited within terminal buds and which tunnel the interior of the twigs.

tolerance In ecology, the ability of an organism to endure under certain environmental conditions, as shade, light, drought, and so forth.

tundra Biome of subarctic regions having permanently frozen subsoil and a low vegetation of lichens, dwarf hardy herbs, and shrubs.

tylosis(es) In wood, outgrowth in a vessel wall from an adjacent cell, partially or completely blocking the opening.

vascular Tissues in plants that conduct food and nutrients, usually xylem and phloem in trees.

vegetative propagation Regeneration of plants by asexual methods.

veneer Thin sheet of wood of uniform thickness.

virgin forest Any forest not modified by humans or their domesticated livestock. Usually a mature or overmature forest uninfluenced by logging, although a young forest naturally regenerated following wildfire is also virgin.

white pine weevil An insect of the genus *Pissodes* that attacks buds and inner-bark of the leading shoot of white pine trees, resulting in deformed stems.

wildling Seedling naturally reproduced in a forest and transplanted for reforestation.

wilting-point, permanent Condition at which plants cannot recover turgidity, even when water is added to the soil.

windbreak A strip of trees or shrubs that provides protection for farmsteads from wind and cold. See *shelterbelt*.

windthrow Trees uprooted by wind.

xeric For soils and sites, meaning dry or adapted to dry conditions.

SCIENTIFIC NAMES OF TREES MENTIONED IN THE TEXT

Aspen, quaking *Populus tremuloides*
Baldcypress, southern *Taxodium distichum*
Basswood, American *Tilia americana*
Beech *Fagus grandifolia*
Fagus americana
Birch, gray *Betula populifolia*
Birch, paper *Betula papyrifera*
Birch, white *Betula alba*
Cherry, wild black *Prunus serotina*
Chestnut, American *Castanea dentata*
Cottonwood, eastern *Populus deltoides*
Dogwood, flowering *Cornus florida*
Douglas-fir *Pseudotsuga menziesii*
Elm, American *Ulmus americana*
Fir, balsam *Abies balsamea*
Fir, white *Abies concolor*
Hemlock, eastern *Tsuga canadensis*
Hemlock, western *Tsuga heterophylla*
Hickory *Carya* spp.
Holly, American *Ilex opaca*
Honeylocust *Gleditsia triacanthos*
Locust, black Robinia pseudoacacia
Maple, red *Acer rubrum*
Maple, sugar *Acer saccharum*
Oak, red (group) *Quercus* (Erythrobalanus) spp.
Oak, white (group) *Quercus* (Leucobalanus) spp.
Pine, eastern white *Pinus strobus*
Pine, hard *Pinus* spp.
Pine, jack *Pinus banksiana*
Pine, loblolly *Pinus taeda*
Pine, lodgepole *Pinus contorta*
Pine, longleaf *Pinus palustris*
Pine, pitch *Pinus rigida*
Pine, pond *Pinus serotina*
Pine, ponderosa *Pinus ponderosa*
Pine, Monterey *Pinus radiata*
Pine, red *Pinus resinosa*
Pine, Scotch (Scot's) *Pinus sylvestris*
Pine, shortleaf *Pinus echinata*
Pine, slash *Pinus elliottii*
Pine, soft *Pinus* spp.
Pine, southern *Pinus* spp.
Pine, sugar *Pinus lambertiana*
Pine, western white *Pinus monticola*
Redcedar, eastern *Juniperus virginiana*
Redcedar, western *Thuja plicata*
Redwood *Sequoia sempervirens*
Spruce, black *Picea mariana*
Spruce, Norway *Picea excelsa*
Spruce, Sitka *Picea sitchensis*
Spruce, white Picea glauca
Sweetgum *Liquidambar styraciflua*
Sycamore, American *Platanus occidentalis*
Walnut, black *Juglans nigra*
White-cedar, Atlantic *Chamaecyparis thyoides*
White-cedar, northern *Thuja occidentalis*
Willow, black *Salix nigra*
Yellow-poplar *Liriodendron tulipifera*

Index